DEALING WITH CLIMATE CHANGE

Setting a Global Agenda for Mitigation and Adaptation

Edited by
R K Pachauri

teri

The Energy and Resources Institute

First reprint 2011
Second reprint 2014
Third reprint 2015
Fourth reprint 2016

ISBN 978-81-7993-277-3

Suggested citation
Pachauri, R. K. 2010. *Dealing with Climate Change: Setting a Global Agenda for Mitigation and Adaptation*. New Delhi: TERI

Published by
TERI Press
The Energy and Resources Institute (TERI)
TERI Press
Darbari Seth Block
IHC Complex, Lodhi Road
New Delhi – 110 003
India

Tel. 2468 2100 or 4150 4900
Fax 2468 2144 or 2468 2145
India +91 • Delhi (0) 11
E-mail teripress@teri.res.in
Web www.teriin.org

Printed in India

CONTENTS

Foreword

How did we get here and where do we need to go?

The habitability of the earth depends on the greenhouse gases that blanket our planet and ensure that it is not a cold rock but a warm and variable surface that can sustain life based on photosynthesis. The changes in climate, brought about, until modern times, mainly by natural causes like changes in the earth's orbit around the sun, variations in solar and volcanic activity, have also been a factor in the continuous variability of ecosystems that has stimulated the evolutionary development of life. Hence the source of our concern is neither the greenhouse effect nor the variability of climate. It is the evidence of human impact, the pace of change, and the capacity of human systems to adapt.

The risks of climate change that now confront us are largely of our own making. Moreover, the pace and extent of change now predicted by scientists exceed anything that human societies have had to cope with in the past. These risks are not a threat to the earth as such. They are a threat to us humans and to other species. If we fail to prevent runaway climate change, the earth will not disappear. We will.

The human impact on climate has been known since 1896, when the Swedish scientist Arrhenius put some numbers to a theory first propounded by Joseph Fourier in 1824. But a serious policy discussion first emerged in international discussions at the First World Climate Conference in 1979. Initially the international response focused on cooperative scientific work, a strand of activity that gathered pace with the establishment of the Intergovernmental Panel on Climate Change (IPCC) in 1988. In parallel, a policy dialogue began at the Second World Climate Conference held in Geneva in November 1990, when the Rio Earth Summit was in the early stages of preparation.

The scientific consensus on the magnitude of change which could be expected, the links with human impact, and the overall impact on ecosystems took time to evolve. In fact, a definitive statement that observed that likely changes were due to human actions and were, therefore, controllable by modifying these actions did not come till the *Third Assessment Report* issued in 2001. The same was asserted forcefully only in the *Fourth Assessment Report* issued in 2007, produced by the IPCC under the chairmanship of Dr R K Pachauri.

The political pressure for action pre-dates the scientific consensus. It came from some developing countries that were going to be severely affected by the one impact that was clear—the rise in the sea level. The then Presidents of Bangladesh and the Maldives raised the issue at the Commonwealth Summit of 1988 and later, at the United Nations. A Commonwealth Expert Group (of which I was a member) was set up and helped to mobilize concern and support for a global process at the governmental level. The growing concern of climate scientists and lobbying by vocal environmental groups led to a commitment to negotiate a framework convention to be ready for signature before the Rio Summit in June 1992.

The Rio Earth Summit and the Climate Convention were both launched during that narrow window of opportunity, when the Cold War was at an end, multilateralism was in the ascendant, with the successful UN sanctioned action in 1991 to undo Iraq's invasion of Kuwait; South Africa was being liberated from apartheid; and the Oslo process was under way, leading to an agreement in 1993 between the Palestine Liberation Organization (PLO) and Israel. As one involved in managing the diplomatic process for Rio, I can say that there were few spillovers from other areas of conflict into the Rio or the climate negotiating process. The main spillover was from North-South tensions on trade and aid, which affected the Rio process more than the climate negotiations.

The Convention that was opened for signature at Rio was merely a framework that did not impose any binding obligations of emission reductions on the parties. But it did enunciate some aspirations and basic principles like 'common but differentiated responsibility'. The attitude of the United States was a critical factor in limiting the scope of the agreement. The US, which hosted the first negotiating meeting for the Climate Convention, was, at this time, under the Bush administration, heavily influenced by skeptics like John Sununu, the then White House Chief of Staff, and by powerful oil industry interests.

The main demand for strong action came from small island countries for the very understandable reason that they were singularly vulnerable to the consequences of global warming with some support from Western Europe, where a combination of North Sea gas and nuclear power facilitated low-carbon growth. The pressure built up, and soon after the Convention entered into force in March 1994, the process of developing a binding protocol began, culminating in the Kyoto Protocol in which the industrial countries accepted binding obligations on emission reductions by 2008–12 with some innovative flexibility mechanisms. In an indirect way, developing countries were brought into the mitigation effort by the Clean Development Mechanism (CDM), which evolved from a Brazilian proposal and allowed industrial countries to fulfil their obligations by undertaking emission reducing activities in developing countries.

In some ways, Kyoto was an experiment rather than a fully worked out mitigation plan. The numbers negotiated there for emissions reduction were not based on any assessment of the magnitude of mitigation required. Nor was the distribution of mitigation effort between the industrial countries based on any assessment of past liability or current ability. The targets set were more or less arbitrary. As one who observed the process from the management end, I believe that certain targets were accepted, even when they seemed out of sync with others, to ensure acceptance by holdouts who played hardball. .

Kyoto was a bazaar bargain, and even this remained incomplete when, with the election of a new administration, the US backtracked on its commitment. As I have stated elsewhere: 'In terms of the classical narrative of the tragedy of the commons, what happened in the climate negotiations was that there was a general agreement that there are too many cows in the field. Then, instead of working out how many there should be, the agreement was that those of us who have the largest herds will withdraw some cows. Of course, the biggest cattle owner has simply refused to go along, saying that he will teach his cows to behave better.'

The Kyoto Protocol came into force on 16 February 2005, more than seven years after it was adopted, which is an indication of how much foot-dragging there was once the US dropped out of the agreement. The actual performance by most of those who did ratify has been significantly short of what they had promised, and this has called into question the usefulness of a process where multilaterally agreed upon caps on emissions are laboriously negotiated. But the cap on emissions is essential if cross-border trade in carbon emission credits is to be promoted. Carbon trading matters because it creates a market for carbon efficiency and instills carbon consciousness at the point where it really matters—in the investment and production decisions of enterprises. The purchase of carbon emission credits from developing country partners in the CDM and the trading of emission allowances between industrial country parties subject to caps have helped to promote greater carbon efficiency at the margin.

The end of the first period of the Kyoto Protocol meant that a negotiating process for the next phase of the protocol had to be launched. But this could not be done without bringing the US into the tent. Moreover, there were growing concerns about the rising emissions from China, India and other developing countries. This led to a two-track process in the Bali mandate, with one track dealing with the commitments under the Kyoto Protocol and the second track dealing with what was called Long Term Cooperation (LCA). The mandate for the LCA called for a long-term vision and a global goal for emissions reduction, though no specific target was stated and there was only a preambular reference to the need for deep cuts. The Annex I industrial countries were required to undertake commitments or actions that are 'measurable, reportable and verifiable'; but there was some ambiguity on whether these have to be multilaterally agreed upon. Mitigation actions by developing countries were tied to the provision of finance and technology. The mandate also required agreements on a vaguely defined sectoral approach, on credits for deforestation avoided, on finance and technology transfer and, most importantly, for meeting the costs of adaptation that will be incurred by countries.

The Copenhagen process ran into difficulties because of incompatible expectations. Europe wanted that the legal form of its commitment and that of the US was similar. The US refused to join the Kyoto Protocol, the continuation of which is implicit in the Bali mandate, and wanted China and India to be put on the same legal basis and take on similar commitments. China and India, for their part, were firmly wedded to the United Nations Framework Convention on Climate Change (UNFCCC) principles, which in their view did not require them to take on obligations comparable to those for the Annex I countries. At the same time, the demand for strong action from the vulnerable small island states and African countries became more insistent. The alarms rung by IPCC's *Fourth Assessment Report* became louder.

The actual outcome at Copenhagen is the outcome of these incompatible expectations. The Copenhagen Accord is a statement of intent, which still requires a further process of negotiation. The endgame at Copenhagen showed that a global agreement on climate depends on the willingness of two key players, USA and China, to offer significant concessions. The other larger players acquire some influence because one or the other of the big two needed their support. The people who were left out of the endgame altogether were the smaller countries, which include many of the most vulnerable ones. It may be premature to mourn the end of the multilateral climate process. But the fact is that the UNFCCC process is badly in need of repair. We must also recognize that much of the action on climate will now come as much from unilateral action by countries under public pressure and from bilateral and plurilateral agreements as from the UNFCCC process.

The contentious issue now is not the science, though there has been some criticism of the IPCC lately. The work of the IPCC, and more particularly its *Fourth Assessment Report*, has convinced most decision-makers that the time for action has come and that we can no longer say that we need to study the problem a little further before doing anything. The economic case for action has been made by the IPCC and the *Stern Review of the Economics of Climate Change.* The real challenge is the lack of agreement on how the principle of 'common but differentiated responsibility' should be implemented. In determining who should do what and when, what is the role of culpability as measured by past cumulative contributions, which determines the present risk of climate change? Should mitigation obligations be specified in terms of targets for annual emissions or on some principle of fair sharing of cumulative emissions in the decades to come? We do not have even the beginnings of a global consensus on this issue of burden sharing.

A global agreement on climate change may be a long time in coming. But the processes of climate change will not wait while we wrangle. Therefore, coping with the risks of climate change requires that we look beyond the negotiating process for the Climate Convention and mainstream climate concerns into the decision-making process on investment, production, and consumption decisions at every level.

In a developing country like India, the challenge for development planners is to reconcile climate action and growth. At the sectoral level, the challenge is greatest in two areas—energy and water resources.

Energy poverty is a reality for most Indians. Eliminating this is as important a challenge as climate change. And so is energy security—ensuring reliable and adequate supplies of a resource that is central to every production process. How do we combine these development compulsions with carbon emission mitigation, which will become necessary as carbon constraints start to bite through quotas, prices or trade measures?

A case in point is renewable energy development. In fact, in India, this programme developed as a response to concerns about energy poverty in rural and remote areas. But it clearly also serves the purposes of mitigation. This is what mainstreaming means. Climate action is not an add-on scheme to other development programmes but is factored in as an objective along with others in the design of projects and programmes.

Water availability is the main, but not the only, route through which adaptation pressures will become manifest. We have to plan water resource projects for agriculture, hydropower, drinking water and industry by adjusting historical hydrological data for potential changes in precipitation, river flow, and groundwater recharge. Even the increased risk of natural disasters is in some ways related to hydrological changes. Ensuring fairness in access to water and efficiency in its utilization is also the key to equitable agricultural growth and rural development. Hence, a water policy redesigned to cope with the impact of climate change can become a tool for sustainable development.

We have to recognize the synergies between mitigation, adaptation, and poverty reduction activities. Forest policy, for instance, has to integrate considerations about changing growing conditions (adaptation), afforestation as a way of sequestering carbon (mitigation) and meeting fuel needs (energy poverty reduction). In fact forests should be seen as a public utility which, besides providing these benefits, also plays a role in soil conservation, hydrological management, biodiversity conservation and so on.

Low-carbon growth to mitigate climate risks and adaptation to the climate changes that are unavoidable will involve adjustments in techniques of production and consumption that are substantial. These changes will require what Schumpeter called 'creative destruction'. New technologies and business models will be required in energy, transportation, housing and habitat, agriculture and water use. For this to happen, policy must stop protecting enterprises that have become unprofitable because of rising carbon costs or changing climatic conditions. Unfortunately, governments are far more prone to give in to pressures from old obsolete enterprises than to support new ventures that are experimenting with the needs of the future

The corporate sector, for its part, must see action on climate change not just as a corporate social responsibility opportunity but as something which needs to be integrated into core corporate strategy. They must anticipate and innovate. When they invest in plants and facilities that will last decades they must adapt to the likely impact of climate change on local weather and hydrology, sea-level, biotic resources, health threats, etc. They must also factor in the potential carbon constraints they will face in the form of direct quotas or carbon taxes. And they must innovate to meet these likely compulsions for mitigation and adaptation.

The great challenge is to reach the mass of small producers and consumers. Raising awareness helps but is not enough. Individuals have to be influenced through prices and the range of technical options available to them to meet their production and consumption needs. This is the role of carbon pricing, either through credits for savings relative to a norm or through carbon taxes. Carbon prices make the carbon constraint bite at the level of micro-decisions by individuals.

The financial and economic crisis that has hit us in 2008 can become an opportunity for some of the actions required to cope with climate change. There is general agreement that pump priming with public spending is going to be necessary to revive the processes of economic growth. Can we not use this opportunity to promote climate action with large-scale public spending on forestry, energy efficiency, water conservation, public transportation, housing and habitat renewal and similar measures?

Global warming is the most important existential threat that humanity faces at the moment. How we cope with it will be a measure of whether nature has succeeded in creating a thinking species. This book is an attempt to assist nature in this endeavour. The authoritative contributions in the volume lay out the issues, the options, and the prospects. Read it carefully, for the future of our children and their children and others beyond depends on our understanding and commitment, individually and collectively.

Nitin Desai
January 2010

Acknowledgements

We gratefully acknowledge the distinguished contributing authors who have agreed to be part of this publication.

Our special thanks are due to Dr R K Pachauri, without whose vision and guidance, this publication would not have been possible. We thank Dr Prodipto Ghosh, Ms Suruchi Bhadwal, Dr Vivek Kumar, and Dr Sanjay Vashisht for their valuable inputs and involvement, especially in the conceptualization and development of the contents.

Thanks are also due to colleagues in TERI Press—Madhu Singh Sirohi for her guidance; Arshi Ahmad for editing and coordinating the process; Richa Sharma for editing; Roshni Sengupta for coordinating with the contributors; R Ajith Kumar for typesetting and designing; R K Joshi for the diagrams and graphics; Priyabrata Roy Chowdhury for the cover design; and T Radhakrishnan for facilitating the production of the book.

1

The scientific assessment of climate change: new findings

R K Pachauri*

The *Fourth Assessment Report* (AR4) of the Intergovernmental Panel on Climate Change (IPCC) revealed a great deal of new scientific knowledge, which carries learning with a special value in a world where climate change has now come to occupy centrestage on contemporary global issues. But what has also been a major revelation is the learning that the world has been party to since the report itself was published. The major revelation of the period since November 2007, when the *Synthesis Report* of the last assessment was released, is the importance of information dissemination in a field that clearly has a major bearing on policies related to growth and development and every sectoral activity pursued by human society. Indeed, the IPCC has been motivated in its actions by creating reports that are policy relevant without being policy prescriptive. But what has been a remarkable advance in our understanding of the role of the IPCC has been the importance of spreading the major findings of the Panel's reports in order to establish the IPCC's policy relevance.

What has emerged from the experience since November 2007 is the fact that in democratic societies in particular, policy changes can only come about if the public desires them, and governments, therefore, sit up and take notice of what the public demands only when such a manifestation is widespread and strong. Such a shift in public perceptions and priorities, however, would come about only if they are

* Chairman, Intergovernmental Panel on Climate Change (IPCC); Director-Gerneral, The Energy and Resources Institute (TERI), New Delhi

informed of the rationale behind the choices that confront them. The scientific assessment of the IPCC, therefore, has to be translated into messages that the public can absorb and act on, creating a relevance on policy changes that leaders and governments will then accept and implement. Clearly, the ability and capacity of the IPCC to spread the message of its findings is limited, and therefore, there is a need for partners to work in the space of information dissemination, which the IPCC is neither equipped nor qualified to occupy arising out its own assessments.

The choice between action and inaction

The Energy and Resources Institute (TERI) has been engaged in research on climate change since 1988, and its very first piece of work in the field was carried out in preparation for a major conference that was organized in New Delhi in January 1989 with the title 'Global warming and climate change: perspectives from developing countries'. This conference was attended by leading scientists of the world, who for the first time focused on the specific conditions in the developing world and how considerations of equity, efficiency, and burden sharing needed to be evaluated in defining the role of the developing countries in meeting the global challenge of climate change, which has essentially been created by emissions that have taken place overwhelmingly in the developed world. Soon after the conference itself, TERI established a Centre for Global Environmental Research, the major focus of which was on climate change. This Centre not only conducted research but also ensured adequate attention to the dissemination of its results and the creation of awareness, which took place not only through spreading the message to the public and schoolchildren in particular, but also in advising governments and multilateral organizations on the basis of analysis that was carried out on a regular basis. Today, the world is clearly situated at the crossroads with choices that might tempt us to continue with business as usual, unwittingly creating a much greater and growing degree of climate change, the impacts of which will become progressively negative. The other choice is for us to evaluate the extent to which we must put in place adaptation measures and implement mitigation of greenhouse gas (GHG) emissions, so that the future does not impact on human society or other living species in a manner that creates unbearably negative or irreversible changes.

One guiding consideration in scientific assessment of climate change arises from Article 2 of the UN Framework Convention on Climate Change (UNFCCC), which clearly states *'The ultimate objective of this Convention and any related legal instruments that the Conference of the Parties may adopt is to achieve, in accordance with the relevant provisions of the Convention, stabilization of greenhouse gas concentrations in the atmosphere at a level that would prevent dangerous anthropogenic interference with the climate system. Such a level should be achieved within a time-frame sufficient to allow ecosystems to adapt naturally to climate change, to ensure that food production is not threatened and to enable economic development to proceed in a sustainable manner.'*

Dangerous anthropogenic interference: the overwhelming evidence

The definition of what constitutes dangerous anthropogenic interference with the climate system is a matter that cannot be defined by science. It is an issue that involves value judgments that go beyond scientific facts and knowledge, but such a determination has to be based on science, and perhaps enabled by science. This means that the impacts of climate change cannot be evaluated on the basis of some vague average values that may define global conditions in general, but essentially these should relate to the impacts of climate change in perhaps the most vulnerable regions in the world. Clearly, what we need to do then is to see that we identify the most vulnerable regions, hypothetically speaking, and then define the conditions of climate change that would push such a location into the category of dangerous. Such an exercise can be problematic in practice both in respect of time and space.

Perhaps it is for this reason that policy-makers and, most certainly, negotiators working on the development of an agreement, both at the stage of drafting the Kyoto Protocol and an agreement for Copenhagen, have shied away from identifying the precise community or location that would be the first to face the danger from the impacts of climate change and defining what these impacts might actually be. There is, therefore, no reason to expect that the global community will come up with precision in applying knowledge to give substance to Article 2 of the UNFCCC. But we certainly need to map out the space on the global map that could be created as a result of unmitigated climate change if we continue to remain inactive in tackling the problem.

TERI has attempted to distil the information from the AR4 of the IPCC and recent developments, on the basis of which the public can be informed about where the world stands on the basis of existing knowledge in dealing with this enormous and unprecedented challenge. Consequently, it would be useful to look out where we are with scientific knowledge on the basis of work carried out by the IPCC and within the framework of what would be intelligible to the general public at large. First, we must understand the powerful findings that come out of the observational evidence that clearly provides substance to the IPCC's findings: '*Warming of the climate system is unequivocal, as is now evident from observations of increases in global average air and ocean temperatures, widespread melting of snow and ice, and rising global average sea level.*' The summary of what has happened since the beginning of industrialization is best provided by Figure 1, reproduced from the IPCC's AR4.

What is significant is the fact that since the beginning of the industrial age, while global average surface temperature has fluctuated over the initial 100 years or so, the last five or six decades have clearly been marked by a clear trend of rising temperatures. This is clearly the result of human actions, which have contributed to a substantial increase in the concentration of GHGs in the atmosphere. Hence, while the previous fluctuations in temperature were the combined result of natural as well as human factors, the second half of the 20th century and the early part of this century have clearly seen the dominance of human actions over natural factors in determining temperature increase. Figure 1 also shows an increase in sea level, which during the 20th century amounted to around 17 cm. Correspondingly, northern hemisphere ice cover has gone down significantly during the second half of the 20th century.

Contributing to these changes are largely human actions, which have resulted in a substantial increase in global GHG emissions since pre-industrial times, with an increase of 70% between 1970 and 2004. This massive increase clearly highlights the ineffective nature of global action, which could have been expected to have had at least some marginal impact, particularly since the UNFCCC came into existence in 1992 and the Kyoto Protocol was ratified and came into official effectiveness in 2005. Human actions indicate a major aberration and departure not only within a time scale of the future decades but as a matter of fact, as the AR4 clearly observed: '*Global atmospheric*

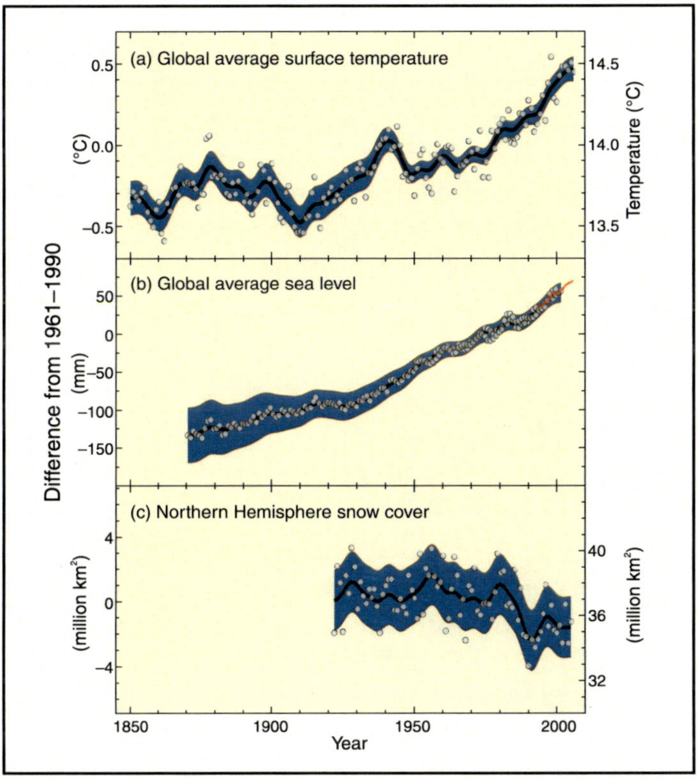

Figure 1 Observed changes in (a) global average surface temperature; (b) global average sea level from tide gauge (blue) and satellite (red) data and (c) Northern Hemisphere snow cover for March-April. All differences are relative to corresponding averages for the period 1961-1990. Smoothed curves represent decadal averaged values while circles show yearly values. The shaded areas are the uncertainty intervals estimated from a comprehensive analysis of known uncertainties (a and b) and from the time series (c).

Source IPCC Synthesis Report Figure SPM.1

concentrations of CO_2*, methane (*CH_4*) and nitrous oxide (*N_2O*) have increased markedly as a result of human activities since 1750 and now far exceed pre-industrial values determined from ice cores spanning many thousands of years.'*

It is also interesting to highlight the fact that during the past 50 years, the sum total of solar and volcanic activities would likely have produced cooling. The actual observed patterns of warming and changes are simulated, as it happens only by models that include the effects of human actions. Consequently, those who doubt the effect of the human hand in causing climate change must look at the evidence provided by the sophisticated and highly reliable models that are being used currently to simulate past actions with the inclusion of both natural as well as anthropogenic forcings. There should, therefore, be no cause for doubt that observed climate change is the result of both natural, as well as anthropogenic factors, with clearly anthropogenic forcings dominating what has happened in the last five or six decades. The importance of this finding is clearly brought out by Figure 2.

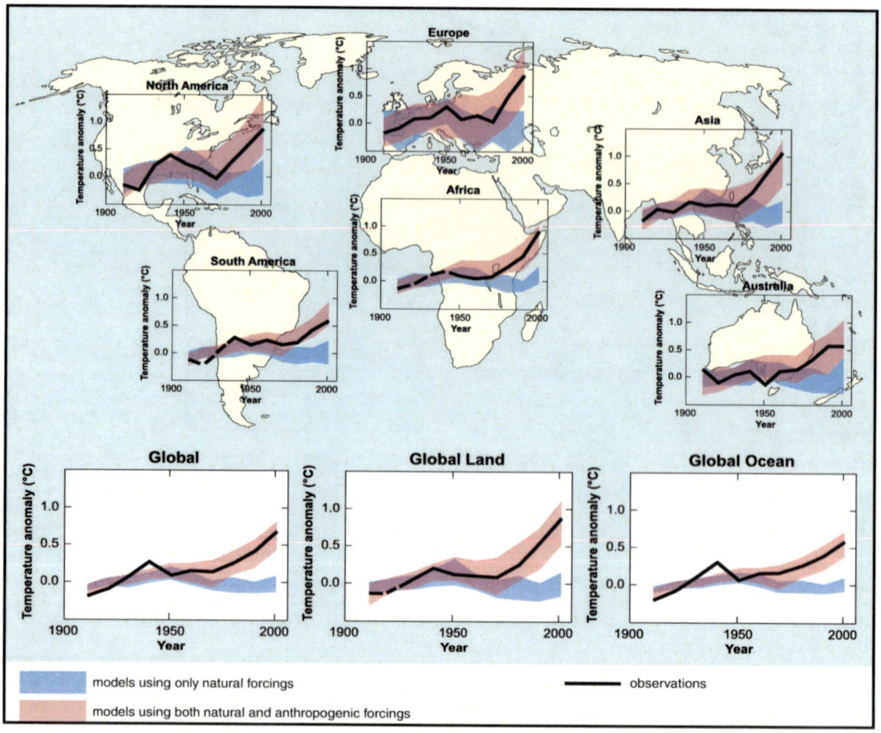

Figure 2 Comparison of observed continental- and global-scale changes in surface temperature with results simulated by climate models using either natural or both natural and anthropogenic forcings. Decadal averages of observations are shown for the period 1906-2005 (black line) plotted against the centre of the decade and relative to the corresponding average for the period 1901-1950. Lines are dashed where spatial coverage is less than 50%. Blue shaded bands show the 5% to 95% range for 19 simulations from five climate models using only the natural forcings due to solar activity and volcanoes. Red shaded bands show the 5% to 95% range for 58 simulations from 14 climate models using both natural and anthropogenic forcings.
Source IPCC Synthesis Report Figure SPM.4

Future projections

The IPCC has carried out projections of the future temperature changes based on a range of plausible scenarios. These are brought out by Figure 3.

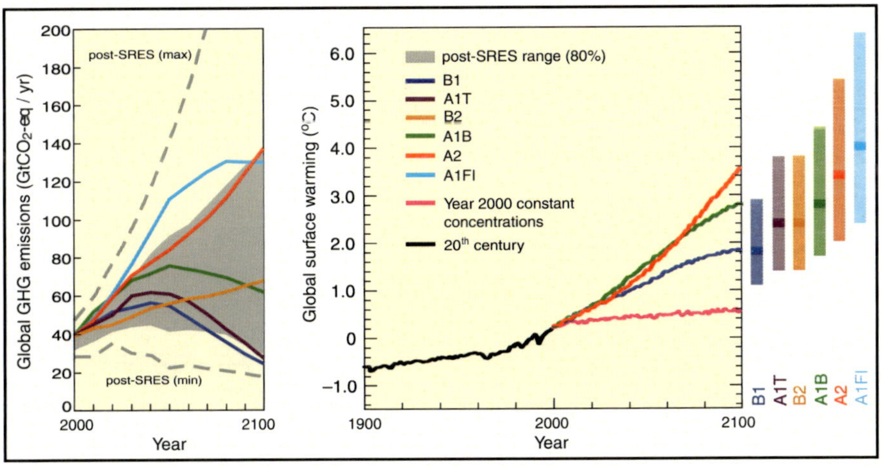

Figure 3 Left panel: Global GHG emissions (in $GtCO_2$-eq) in the absence of climate policies: six illustrative SRES marker scenarios (coloured lines) and the 80th percentile range of recent scenarios published since SRES (post-SRES) (gray shaded area). Dashed lines show the full range of post-SRES scenarios. The emissions include CO_2, CH_4, N_2O and F-gases. Right panel: Solid lines are multi-model global averages of surface warming for scenarios A2, A1B and B1, shown as continuations of the 20th-century simulations. These projections also take into account emissions of short-lived GHGs and aerosols. The pink line is not a scenario, but is for Atmosphere-Ocean General Circulation Model (AOGCM) simulations where atmospheric concentrations are held constant at year 2000 values. The bars at the right of the figure indicate the best estimate (solid line within each bar) and the likely range assessed for the six SRES marker scenarios at 2090-2099. All temperatures are relative to the period 1980–99.
Source IPCC Synthesis Report Figure SPM.5

In essence, the projected increase in temperature by the end of the 21st century would likely be within a range of 1.1 °C to 6.4 °C. In order to help in focusing on what may seem a clear target for human society to work with, the IPCC pointed to a best scenario of 1.8°C at the lower end of the range and 4 °C at the upper end. It is important to understand that if we were to take even the lower end best estimate of 1.8 °C, we would reach a total increase of over 2.5 °C, given that the average increase during the 20th century has been 0.74 °C. The impacts of climate change with this overall increase in temperature are likely to be serious for some key sectors and ecosystems. A useful understanding of the intensity and spread of these impacts is represented in Figure 4.

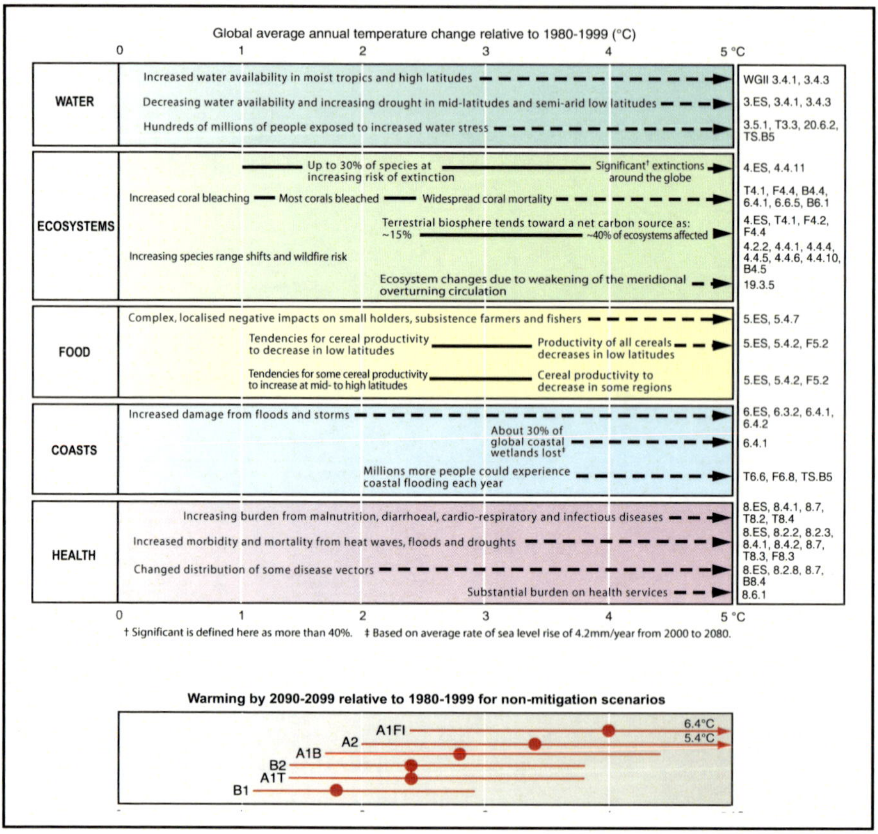

Figure 4 Examples of impacts associated with projected global average surface warming. Upper panel: Illustrative examples of global impacts projected for climate changes (and sea level and atmospheric CO_2 where relevant) associated with different amounts of increase in global average surface temperature in the 21st century. The black lines link impacts; broken-line arrows indicate impacts continuing with increasing temperature. Entries are placed so that the left-hand side of text indicates the approximate level of warming that is associated with the onset of a given impact. Quantitative entries for water scarcity and flooding represent the additional impacts of climate change relative to the conditions projected across the range of SRES scenarios A1FI, A2, B1 and B2. Adaptation to climate change is not included in these estimations. Confidence levels for all statements are high. Lower panel: Dots and bars indicate the best estimate and likely ranges of warming assessed for the six SRES marker scenarios for 2090-2099 relative to 1980-1999.

Source IPCC Synthesis Report Figure SPM.7

In some parts of the world, the impacts would clearly be far more serious than in other regions. For instance, the IPCC has identified the mega-deltas, which include cities like Shanghai, Dhaka, and Kolkata, as extremely vulnerable to coastal flooding. These are locations with high population density and substantial infrastructure, and given their low-lying nature and vulnerability to coastal flooding, the extent of damage and loss that could take place as a result of the current climate-related events could be very high. Other impacts of climate change include exacerbation of existing stresses such as in parts of Africa. It has been projected, for instance, that by 2020 there would be 75–250 million people living under conditions of water stress in Africa as a result of climate change. It must be kept in mind that there are communities and countries on that continent which are already under high levels of stress connected with poverty, malnutrition, widespread disease, and existing levels of water stress. In addition, as early as 2020 itself, some parts of Africa are likely to see a decline in agricultural yields for some crops by about 50%.

The Arctic region is also particularly vulnerable, with a substantial loss of see ice. This region in fact is warming at twice the rate of the rest of the globe. Particularly significant would be the possibility of abrupt and irreversible changes that could take place as a result of climate change. For instance, if any part of the Greenland or West Antarctica ice sheet were to collapse, the world would possibly face sea-level rise of several metres. This could result not only in large-scale destruction but also the problem of 700 million environmental refugees in different parts of the world.

Against this prospect of serious impacts of climate change based on the scenarios of unmitigated emissions growth in the future, two imperatives emerge very clearly. First, given the inertia in the system, and the fact that temperature increases and other changes in climate are inevitable over the next few decades, adaptation to the impacts of climate change has to be taken in hand with a sense of urgency. Second, the impacts would over a period of time reach a magnitude that would clearly exceed the capacity of communities and countries to be able to adapt. Consequently, mitigation of emissions has to be pursued with a sense of urgency and at a level that would bring about stabilization of the earth's climate.

As it happens, the cost of mitigation is so attractive and the co-benefits associated with it so important that a path of stringent mitigation for global society is clearly an extremely attractive approach on scientific and economic grounds. For instance, to stabilize temperature increase between 2.0 °C and 2.4 °C, the extent of mitigation required would impose a cost of no more than 3% of the global gross domestic product (GDP) in the year 2030. If, however, we add the co-benefits from such a mitigation pathway – higher energy security, lower levels of air pollution and, therefore, commensurate health benefits, higher levels of employment and the benefits of avoiding negative impacts of lowered agricultural productivity – the world would incur in the aggregate negative costs related to such a path of mitigation. The IPCC AR4 also clearly concluded,

> *'Many impacts can be reduced, delayed or avoided by mitigation. Mitigation efforts and investments over the next two to three decades will have a large impact on opportunities to achieve lower stabilization levels. Delayed emission reductions significantly constrain the opportunities to achieve lower stabilization levels and increase the risk of more severe climate change impacts.'*

Another important conclusion that the AR4 provides was the fact that if the world wants to stabilize temperature increase between 2.0 °C and 2.4 °C, we have very little time to embark on the preferred trajectory of mitigation. In essence, global emissions will have to peak no later than 2015. This clearly emphasizes the importance of early action and the significance of an agreement that provides substance to the Bali Road Map and ensures firm commitments for reduction in GHG emissions by 2020. The Bali Road Map drawing on the findings of the IPCC *Synthesis Report* clearly states,

> *'The Conference of the Parties,*
> *Recognizing that deep cuts in global emissions will be required to achieve the ultimate objective of the Convention and emphasizing the urgency*[1] *to address climate change as indicated in the Fourth Assessment Report of the Intergovernmental Panel on Climate Change*

[1] Contribution of Working Group III to the Fourth Assessment Report of the Intergovernmental Panel on Climate Change, Technical Summary, pages 39 and 90, and Chapter 13, page 776.

> *Decides to launch a comprehensive process to enable the full, effective and sustained implementation of the Convention through long-term cooperative action, now, up to and beyond 2012, in order to reach an agreed outcome and adopt a decision at its fifteenth session.'*

Against this background, the decision by world leaders, including President Obama of United States, Prime Minister Lars Løkke Rasmussen of Denmark, and President Hu Jintao of China, among others, not to adopt a binding agreement at Copenhagen, by which the developed countries of the world would reduce GHG emissions is disappointing. In effect, this decision has been postponed to the next Conference of the Parties, to be held in Mexico. While prior developments had been pointing to such an outcome in Copenhagen, the decision by key world leaders to postpone any binding agreement in Copenhagen has been a source of disappointment for those who understand the imperatives of early action to meet the challenge of climate change.

Need for a mass movement on climate

Since 1992, when the UNFCCC came into existence, the global community has really not acted in keeping with expectations that are based on the knowledge and the scientific assessment brought out by the IPCC. In this respect, global action and even initiatives by nation states have proved inadequate and much too late. Such performance by the governments of the world raises questions on whether the nation state is really capable of taking action in adequate measure in the field of climate change. It almost seems that Mahatma Gandhi's approach of creating a mass movement through non-violence and civil disobedience may be the only possible approach for global action in the field. But such a response at the global level would require a 'global Gandhi', who is nowhere in sight today. Several years ago, there was a movement in Europe against a particular oil company, asking for a boycott of its products. Such small splinter movements have surfaced from time to time, but to get a consumer-based global response that could really help to fill up the vacuum created by the inaction of governments is an issue that cannot be treated as a likelihood in the coming years. However, if people across the globe are aroused, perhaps this may be the only means to achieve a major global response that could make a difference towards addressing the problem of climate change, which we see becoming larger and becoming more serious in the coming years. Perhaps the following chapters of this book would help readers across the globe understand not only the nature of the problem but also some of the institutional responses that are required for the world to move on and tackle the challenge adequately and on a timely basis.

It is important for global society to accept and appreciate the inevitability of lifestyle and other changes, so that timely and adequate action can be mounted for stabilizing the earth's climate and ensuring that suitable adaptation measures are taken in hand in response to that component of climate change which has now become inevitable as a result of emissions and increase in concentration of GHGs, which have occurred in the past. The world, therefore, faces a dilemma. On the one hand, there is urgency in responding to the challenge of climate change, so that its impacts do not overwhelm the ability of human society to function on a stable and secure basis. On the other hand, there is a certain social inertia in the system, which favours the continuation of existing lifestyles and patterns of production and consumption that have built up over several decades. As would be expected, continuation of business as usual has the support of a large range of vested interests, who would indulge in widespread advertising and the services of lobbyists to ensure that action is thwarted or delayed for as long as possible. What we are seeing today is perhaps a combination of these factors. The issue really is one that requires individuals going against the tide of social pressures and practices, which have become deeply ingrained in most societies across the globe. Grassroots actions to mitigate climate change on a sizeable scale would require almost a revolutionary fervour, more or less similar to what a leader like Gandhi or his great follower Martin Luther King was able to achieve. We need to ask whether another Gandhi might emerge on the scene to provide direction to the current generation, which is still young and would face the real consequences of climate change. Perhaps, action by civil society, businesses, research and academic institutions and government at all levels would create the conditions for several Gandhis to emerge in due course, and take up this challenge in different parts of the world.

2

Carbon budget proposal

Jiahua Pan and Ying Chen*

With the adoption of the Bali Action Plan at the 13th session of Conference of the Parties (COP-13) of the United Nations Framework Convention on Climate Change (UNFCCC) in December 2007, the Ad hoc Working Group on further commitments of Annex I Parties under the Kyoto Protocol (AWG-KP) and Ad Hoc Working Group on Long-term Cooperative Action under the Convention (AWG-LCA) are taken as the major vehicles in parallel for international climate negotiations, aiming at reaching an agreement on a climate regime beyond 2012. The Principle of Common but Differentiated Responsibilities (CBDR) is an important foundation for building an equitable and sustainable international climate regime beyond 2012. Although equity per se is not a separate topic for negotiation, it is certainly the basis for burden-sharing and the design of implementation mechanisms that are related to mitigation, adaptation, technology, and finance. The core part of the climate architecture lies in differentiation of responsibilities while taking into account national circumstances of individual countries.

There has been a great variety of proposals with different interpretations of the equity principle for burden-sharing and emission entitlements since the start of international climate negotiations in the early 1990s. From a geopolitical point of view, however, there is

* Research Centre for Sustainable Development, Chinese Academy of Social Sciences (CASS/RCSD).

an imbalance of proposals between the South and the North, with the South under-represented. For instance, only three framework proposals are attributable to researchers from the South, out of 43 proposals reviewed in a report by the Pew Center on Climate Change (Bodansky, Chou, and Jorge-Tresolini 2004). As one of the approaches from the developing world, the basic needs approach calls for priority allocation of emissions for basic needs satisfaction by promoting low-carbon development, and discouraging luxurious and wasteful emissions (Pan 2002; Pan, Xianli Zhu, and Chen 2005). This conceptual framework of basic needs is further developed to include a carbon budget for individual earth citizens subject to stabilization of greenhouse gas (GHG) concentration in the atmosphere as the basis for building an equitable and sustainable climate regime (Pan 2008). The Carbon Budget Framework is well grounded and widely accepted from a theoretical point of view, but methodologies are yet to be developed for numerical analysis on the design and implementation of such a regime. The Carbon Budget Proposal (CBP) as developed in the report, therefore, attempts to substantiate the conceptual framework with methodological details for an equitable and measurable allocation of emission entitlements so as to make it a comprehensive design for an international climate regime beyond 2012.

The structure of the report is given as follows. First, based on a brief review and summary of present approaches, the basic principles and design ideas of the CBP are put forward. Key assumptions and methodologies are developed, reflecting requirements for both equity and sustainability. This is followed by the preliminary results and major policy implications. In accordance with the Carbon Budget Framework, elements of climate regime building are addressed. Is CBP in favour of China as the most populous country in the world? This question is also examined in depth. Conclusion is drawn with further discussion on the salient features of this proposal and additional work for its refinement.

Overview of existing proposals for climate change mitigation

Various approaches have been proposed in relation to the design of climate architecture up to and beyond 2012 (Winkler 2008). Examples include continuation of the Kyoto-style for setting a fixed emission reduction target; the application of a universal carbon tax;

technology agreements (Edmonds 1998), Triptych approach extended to the global context (Groenenberg, Phylipsen, and Blok, *et al.* 2001; Den Elzen, Hoehne, Brouns, *et al.* 2007), converging markets (Tangen and Hasselknippe 2004; Victor, House, and Joy 2005), safety-valve approaches (Burtraw 2009), sustainable development policies and measures (SD-PAMs) (Winkler, *et al.* 2002, 2007), and allocation of emissions on a per capita basis (Aslam 2002; Meyer 2000, 2004) The Brazilian proposal advocates an allocation of emissions allowances in relation to the contribution to change in temperature (Brazil 1997; La Rovere, de Macedo, and Baumert 2002; Rosa and Ribeiro 2001; UNFCCC 2002). Among the more recent suggestions are common but differentiated convergence (Höhne, Den Elzen, and Weiss 2006), emissions intensity (Herzog, Baumert, and Pershing 2006; Kim and Baumert 2002; Chung 2007), sector-based Clean Development Mechanism (CDM) (Samaniego and Figueres 2002; Sterk and Wittneben 2006), other sectoral approaches (Ward 2008; Schmidt, Helme, Lee, et al. 2006; Ellis and Baron 2005), Greenhouse Development Rights (GDR) (Baer, Athanasiou, and Kartha. 2007). Still new proposals are emerging.

In contrast to most proposals in the literature, which focus on the production side for international equity, the basic needs approach as developed by Chinese researchers pays attention to emission requirement from consumption side for inter-personal equity (Chen and Pan 2004; Pan 2002, 2008). The concept of a carbon budget is based on two axioms: (a) basic human needs are limited for individuals as a biological entity, but luxurious and wasteful emissions are unlimited; and (b) there exists a geophysical limit due to the finiteness of our planet earth. Interpretations of the equity principle can vary but there must be a common base.

To begin with, first, should equity be discussed among people or between political entities? International negotiation is carried out by representatives of individual countries as political entities. However, international equity does not necessarily mean equity among individuals within a country. It is individual human beings who require, either individually or collectively, basic necessities such as clothing, food, shelter, mobility, and the like. Therefore, emissions should be first allocated for individuals for their survival and basic needs satisfaction.

Second, the key point to promote equity among individuals is to protect the rights of all people living on this planet. If this is the basis for carbon budgeting, should we use the number of people in the past, present or the future? As the current generation inherits the build-up of emissions in the past and determines the size of population in the future, carbon should be budgeted to individuals of the current generation. This would effectively discourage the expansion of population for more carbon budgets as our planet earth is limited. However, migration from one place to another should take along the budget.

Third, equity among people does not merely refer to the fairness in the flux of emissions at an exact moment or points of time in history. Instead, it refers to the stock of emissions in history, at present time and also in the future. It should be evaluated by an accumulative basis covering a designated period of time for completion of a development process. This is because national development is a long-term process and the developed countries started their industrialization at a much earlier time compared with the developing ones. Countries that developed earlier have amassed considerable carbon asset in the past, which is inherited by the current generation with no need to accumulate carbon stock. The developing countries as late comers, however, start their industrialization at a much later point in time in human history. Consequently, these countries accumulated much less carbon assets in the past. Generally, the level of development in developing countries is rather low and basic needs still have to be met. It dictates that in the upcoming industrial process, these developing countries may require more emissions.

Last but not the least, promoting equity between people should take each country's specific circumstances into account and prioritize the basic needs of people. This is because climate, geography, and natural resources endowment could have a direct bearing on carbon emissions for meeting human basic needs. As we know, these natural situations are predetermined and stable and long-term factors that bear little relationship to the development level of a certain country. Evidently, there is a strong case in the budgeting process to make necessary adjustments for accommodating national circumstances.

The current proposals related to emission rights per capita or based on historical emissions (for example, the Brazilian Proposal) could not effectively address the principle of equity. For example, the approach of contraction and convergence (Meyer 2000, 2004) gives full respect to the current situations and strikes for equal emission per capita in a future point of time. However, it fails to take historical, current, and future situations into consideration, thus running the risk of creating greater unfairness (Chen and Pan 2004). Another alternative is the Common but Differentiated Convergence. It highlights a transitional period, in which developed countries are given reduction targets, and most developing countries adopt fixed growth targets. In a second period, global emissions are allocated to countries if their emissions meet a certain per capita threshold. This approach relaxes its restraints on the developing countries at first glance, but it does not acknowledge the rights to and a budget for basic needs emissions by individuals.

The Brazilian Proposal (Brazil 1997) is considered a typical approach, taking into account historical responsibility. This is an equity-based proposal-based not on current but on historical emissions, because GHGs have continuously been piling up since the start of the Industrial Revolution some 200 years ago. This proposal in its original form aims to set historical responsibilities for Annex I countries, and later developed by experts from developed countries for inclusion of all the developing countries (UNFCCC 2002; Rosa and Ribeiro 2001). However, this historic responsibility-based emission allocation approach fails to consider emissions per capita and the realistic needs of each country in different stages of development, as it stresses total national emissions and the polluter-pays principle.

EcoEquity and SEI (Kartha, Athanasiou, Paul Baer, *et al.* 2008) have recently developed a framework of Greenhouse Development Rights (GDR). This holds that only the rich countries should have the obligation and responsibility to reduce GHG emissions. The development needs of the poor countries are protected by the setting of thresholds. By defining the capacity as measured by GDP (purchasing power parity adjusted) and total responsibility (accumulated emissions in history) above the thresholds, the GDR approach allocates emission quota for individual countries. However, this method only takes the historical emissions into consideration and fails to budget emissions for future needs. In addition, there are a few methodological problems such as the conceptualization of threshold values, the calculation of accumulative emissions in history and the source of data, and so on.

The basic line of thinking in this chapter is to formulate a solution through a carbon budget for individuals, with dual goals for meeting both the basic needs for human development and the environmental sustainability.

Key assumptions and methodologies

Carbon dioxide (CO_2) is the major source of GHGs, accounting for 76.7% of global total emissions. CO_2 from fossil fuel combustion has a share of 73.8% over the total CO_2 emissions (IPCC 2007). In budgeting carbon emissions, CO_2 emissions related to fossil fuel are used in this report, since statistics are readily available and are more reliable than those for land use, land use change, and forestry (LULUCF).

The widely used fossil fuels and industrial process CO_2 emissions database from Oak Ridge National Laboratory (CDIAC 2008) provides the most completed time series data of historical emissions before 2005, including most countries and regions. The starting year is a very important parameter for the estimation of global emissions space and accumulative emissions by different countries considering the different historical emissions characteristics among countries. As a result, a comparison is made for 1850, 1900, and 1960 as the starting years respectively.

As for scenarios of future global emissions, global long-term targets could be given in different ways. Some of them are well known, including (a) Kyoto-type emissions targets for stabilizing GHG concentration in the atmosphere (450 ppm for instance), (b) targets to stay below the 2°C threshold as recommended by the European Union (EU), and (3) a convergence target to realize two tonnes of carbon emissions per capita before 2050, as stated in the new Stern Report (Stern 2008). It is true that there is a certain functional relationship between emissions, concentration, and temperature increase but there exists a huge uncertainty for a simple relationship (IPCC 2007). In this report, global emissions reduction target specified at the G8 summit in June 2008 is adopted, indicating a reduction of 50% in 2050 as compared to the 2005 level. The end year of carbon budget is 2050 for assessment in this report.

With the starting year (1900), base year (2005), and the target year (2050) specified, there is a need to design emissions pathways. Relative studies indicate that corresponding to the target to stay below the 2°C threshold, one of the possible emissions pathway will peak in 2017, which is 10% higher than the current level, and then fall rapidly, which will realize 50% emissions reduction by 2050 (Baer 2006). And according to the conclusion drawn in the *Stern Review* (Stern 2006), the next one to two decades will be critical to stabilize the GHG concentration, and certain measures should be taken to inspire global GHG emissions reduction. With reference to projections of global emission in the future by the International Energy Agency (IEA) (2007), two scenarios of global emissions pathways are considered for illustration (Figure 1). Scenario A assumes that global emission level peaks in 2015, at 10% higher than the 2005 level and under Scenario B, global emission level peaks in 2025, at 20% higher than the 2005 level.

Technical calculation and confirmation on increase in atmospheric concentrations and change in global surface temperature under the two scenarios is beyond the scope of this report. Further quantitative analysis is needed to using complicated climate models.

Data about population and energy consumption, degree-days for heating and cooling, geographical factors, and so on are from WRI-CAIT database (WRI-CAIT 2003).

All the countries are divided into two groups, Annex I and Non-Annex I countries. A few typical countries (or group of countries) are selected for detailed investigation, including EU15, EU12, Canada, Japan, Russia, US and Australia. Details are also provided for a few EU member states (France, Germany, Italy, and UK). Representative Non-Annex I countries (Brazil, China, India, Indonesia, Korea,

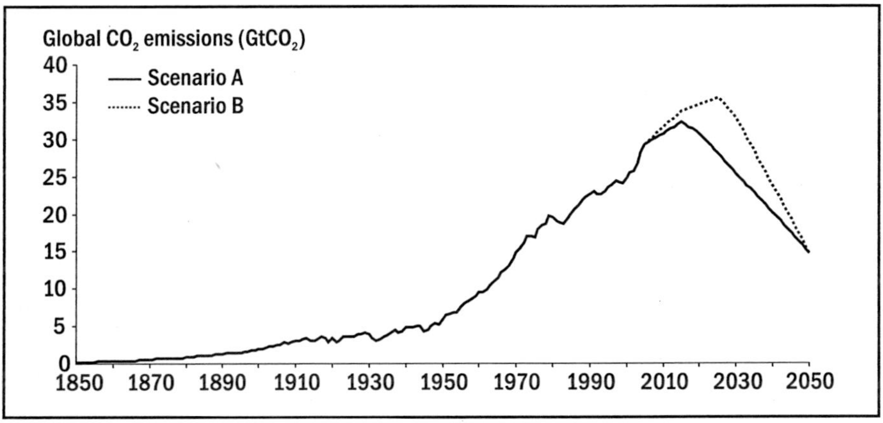

Figure 1 Future emission scenarios
Note Gt – gigatonnes
Source Historical data from CDIAC (2008) and future emission scenarios from author's assumption with reference to IEA (2007).

Mexico, South Africa, Nigeria, Kenya, Zimbabwe, Saudi Arabia, Venezuela) are selected for investigation in this report.

Initial allocation, adjustments and arrangement of transfer payments

Using data and methodology as discussed above, quantitative carbon budget is calculated and the preliminary results are given in this section and their policy implications are discussed.

Specification of global total carbon budget

Global carbon budget under Scenarios A and B, that is, global accumulative carbon emission is calculated, and compared with respect to three starting years all ending in 2050 (Table 1). From Table 1 it is clear that the differences of starting years between 1850 and 1900 are negligible with respect to the budget for global total emissions as historical total emissions before 1900 are rather limited.

Table 1 Estimation of global carbon budget ($GtCO_2$)

Starting year	*Global carbon budget (2050)*		*Historical accumulative emissions (2004)*	*Future carbon budget (2005–50)*	
	Scenario A	*Scenario B*		*Scenario A*	*Scenario B*
1850	2311.1	2472.0	1143.5	1167.6	1328.5
1900	2272.5	2433.4	1104.9		
1960	2019.1	2180.0	851.5		

Note Gt – gigatonnes

Source Authors' calculation based on data from IEA (2007), CDIAC (2008).

Along with growth of the global economy, global emission accelerates year by year led by early industrialized countries. After 1900, the difference between starting years does matter. For example, the gap between 1900 and 1960 reaches 23%. Taking 2050 as the end year, year 1900 is chosen as the starting point. The discrepancy of global future carbon budget is about 14% between Scenario A and Scenario B. The later the peak time, the higher the emission in total budget, the bigger the future emission space, and vice versa.

Initial allocation

Table 2 presents the result of global initial carbon budget as consumed in the past and available for the future up to 2050.

Table 2 Global initial carbon budget (tCO_2/c)

	Total per capita carbon budget (1900 to 2050)		*Historical per capita budget (1900–2004)*	*Future per capita budget (2005–50)*	
	Scenario A	*Scenario B*		*Scenario A*	*Scenario B*
Accumulative emission per capita	352.5	377.5	171.4	181.1	206.1
Annual emission per capita	2.33	2.50	1.63	3.94	4.48

Note tCO_2/c – tonnes of CO_2 per capita
Source Authors' calculation using total budget in Table 1 as divided by total population in 2005.

There are two ways to make the global initial carbon budget during the whole appraisal period, accumulative emission budget per capita and emission budget per capita per year. For instance, the accumulative emission budget per capita is 352.5 tCO_2 from the year 1900 to 2050, while emission budget per capita per year is 2.33 tCO_2 under Scenario A. Historical emission has consumed a total budget at 171.4 tCO_2 in the past 105 years, meaning that the rest of the global budget (about 50% of the whole emission budget) can be used for the remaining 45 years. It is possible to realize the global climate protection goal within the carbon budget if every 'earth citizen' is able to share equally the future emission space, that is, the future emission will not exceed 3.94 tCO_2 per person per year under Scenario A. The budget is rather tight but still feasible. To allocate the global emission entitlement first by the initial carbon budget, as Figure 2 shows, each country's emission budget in aggregate is proportional to the size of their population over global total in the base year (2005 in this analysis). Some countries have larger emission budget due to their high share of population in the base year, such as China and India, while some other countries have smaller emission budget at country level because of the small size of their population, as in the cases of Canada and Australia.

Adjustments

Equal share per capita as the basis for initial carbon budgeting is able to meet the principle of equity by considering historical responsibility and per capital emission. But there are many factors affecting the requirement of carbon budgeting, which require adjustments of the

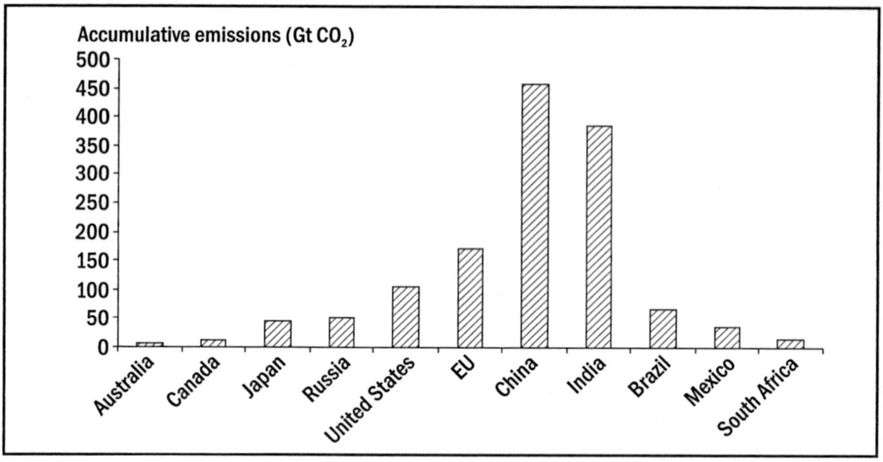

Figure 2 Initial allocation of carbon budget (1900-2050) for Scenario A
Note Gt – gigatonnes
Source Authors' calculation using global initial carbon budget in Table 2 as times population in

initial budgets. Three factors are singled out for the adjustment of budgets, including climatic condition, geographical condition, and resource endowment.

Climatic condition

Climatic condition is an important factor affecting energy demand and carbon emission. In general, heating and cooling are required for the buildings sector only, which accounts for around one-third of the total final energy consumption in developed countries. Energy consumption in heating and cooling is about 50% of the building energy consumption. Other energy consumption is less related to climate conditions, such as cooking, electric appliances, and so on. Therefore, about one-sixth of the global initial carbon budget is assumed to be affected and needs to be adjusted. The climatic adjustment coefficients for different countries are given in Table 3, which is based on population weighted heating and cooling degree days (Baumert and Selman 2003).

Table 3 indicates that countries with cold climate such as Canada, Russia, and Germany would require more heating, while those with hot climate such as India and Indonesia would demand emissions for cooling, assuming that heating and cooling require the same amount of emissions as a result of adjustment coefficient.

Table 3 Climatic adjustment coefficient for selected countries

	Heating degree days	*Population weighted heating degree days*	*Cooling degree days*	*Population weighted cooling degree days*	*Climate adjustment coefficient*
Canada	4493	143 720	171	5485	1.09
France	2478	149 982	241	14 609	0.98
Germany	3252	268 337	122	10 046	1.02
Italy	1838	106 940	600	34 919	0.97
Russia	5235	753 014	197	28 268	1.14
US	2159	633 973	882	259 119	1.00
Australia	828	16 633	839	16 855	0.93
Brazil	118	21 711	2015	370 508	0.95
China	2158	2 797 185	1046	1 355 638	1.01
India	80	86 561	3120	3 368 740	1.01
Indonesia	0	65	3545	771 389	1.03
Mexico	364	37 190	1560	159 217	0.94
South Africa	630	29 219	824	38 218	0.91
Kenya	91	3039	1265	42 333	0.90
Zimbabwe	349	4518	1010	13 066	0.91
Saudi Arabia	311	7010	3136	70 642	1.02
Venezuela	1	33	2381	62 218	0.96

Source Authors' calculation based on data from WRI (2003).

Geographical condition

Geographical condition is one of the factors that influence energy consumption and reveals the fact that carbon emission budget for Russia and Canada would lead to a positive change while that for South Africa, Australia, Mexico, Brazil would mean a minor negative adjustment though small due to their mild climate. As the balance of adjustment has to be neutral, the range of change is between -10% and +14% emissions as mobility requires carbon emissions. In general, energy consumption from the transport sector accounts for around one-third of the total final energy consumption in developed countries. There are two impact factors in the demand of energy and emission in the transportation sector—comfort and average travel distance. In this chapter, the latter factor will only be discussed, which is relevant to geographical condition, supposing that one-sixth of the global initial carbon budget has a close relation to geographical condition and needs to be adjusted. Taking the human activity impact area as adjust factor, we can get the geographical adjustment coefficient at shown in Table 4.

Table 4 Human activity impacted areas and geographical adjustment coefficient for selected countries

	Human activity impact area (1000 km²)	*Human density in human activity impacted area (km² per 1000 people)*	*Total national land area (1000 km²)*	*Adjustment coefficient*
Canada	788	24285.3	9221	1.62
France	456	7485.2	550	1.10
Japan	153	1196.1	365	0.88
Russia	2706	18900.0	16889	1.42
UK	211	3509.6	241	0.96
US	2918	9841.4	9159	1.18
Australia	543	26624.7	7682	1.85
Brazil	2236	11967.5	8457	1.28
China	2804	2149.3	9327	0.91
India	2068	1889.4	2973	0.90
Indonesia	536	2427.9	1812	0.92
Korea (S)	43	884.0	99	0.86
South Africa	497	10595.3	1221	1.24
Nigeria	188	1328.6	924	0.88
Kenya	117	3275.9	580	0.96
Zimbabwe	212	16179.1	391	1.46

Source Authors' calculation based on data from WRI (2003).

We can see from Table 4 that a huge discrepancy exists between national land area and human activity impacted area. It implies that this parameter is more suitable and reasonable to reflect the demand of emission and energy in the transport sector. The result shows that some countries with lower population density will benefit from the adjustment, getting bigger emission budget, such as Australia, Canada, and Russia, while those with higher population density will have smaller emission budget, such as South Korea, Japan, and India. The range of total adjustment varies from -14% to +62%.

Resources endowment

The energy consumption structure of a country is subject to its resource reserves, especially energy resource endowment. But economic power takes an important role in the structure of energy consumption. Sometimes it can change the structure of energy consumption. For instance, Japan can consume petrol oil from other countries though Japan is not rich in resources endowment. As the largest petrol oil consumer in the world, the US has abundant coal reserves and its consumption structure has little connection with its resource endowment. Developing countries have weaker economic strength, and their energy consumption structure will be restricted by national resource endowment. If a country is rich in coal reserve but poor in natural gas reserve, greater share of coal is normally expected in the energy mix despite coal having a high carbon content. This structure of energy consumption is linked to national energy security. It can be reasonable for compensating those countries' emission budget using carbon intensity data. But overcompensation may penalize those who invest more in low-carbon energy. Assuming that half of the global carbon budget is affected and needs to be adjusted, resource endowment coefficient for adjustment is given in Table 5.

Figures in Table 5 show that only a few countries use coal as the main energy source, such as China, India, and South Africa. Higher carbon content per unit of energy in the structure of energy consumption is indicated. While some other countries have lower carbon content, including both developed and developing countries, such as France, Canada, Italy, Brazil, and Mexico, for which carbon budget requires a positive adjustment. The range of total adjustment varies from -40% to +25%.

Overall adjustment

Finally, the above three adjustments have to be combined for calculation of an overall efficient for budget adjustment. As shown in Figure 3, the overall adjustment is in the range of -20%~+78% as these three factors can offset one another to some extent. The geographical element would play the biggest role in the adjustment process among three factors, while climatic element is the smallest.

These natural elements, climate, geography, and resources endowment will play a long-term role in future and can hardly be

Table 5 Energy consumption, carbon content per unit of energy and resource endowment adjustment coefficient for selected countries

	Energy consumption in 2004 (kTOE.)	*Carbon content per unit of energy in 2004 (tCO_2/TOE)*	*Coal in energy mix (%)*	*Adjustment coefficient*
Canada	269 048	1.97	9.9	0.86
France	275 169	1.36	4.8	0.74
Germany	348 036	2.31	25.9	0.92
Italy	184 460	2.44	9.3	0.95
Japan	533 201	2.35	23.5	0.93
Russia	641 532	2.38	15.8	0.93
UK	233 689	2.37	16.8	0.93
US	2 325 887	2.47	24.2	0.95
Brazil	204 847	1.55	6.1	0.78
China	1 609 348	3.13	69	1.08
India	572 851	2.33	54.5	0.93
Korea (S)	213 045	2.21	24.4	0.90
Mexico	165 475	2.41	6.2	0.94
South Africa	131 137	3.17	76.4	1.08
Kenya	16 920	0.63	5.5	0.60
Venezuela	56 158	2.43	0	0.94

Note kTOE – kilotonnes of oil equivalent

Source Authors' calculation based on data from WRI (2003).

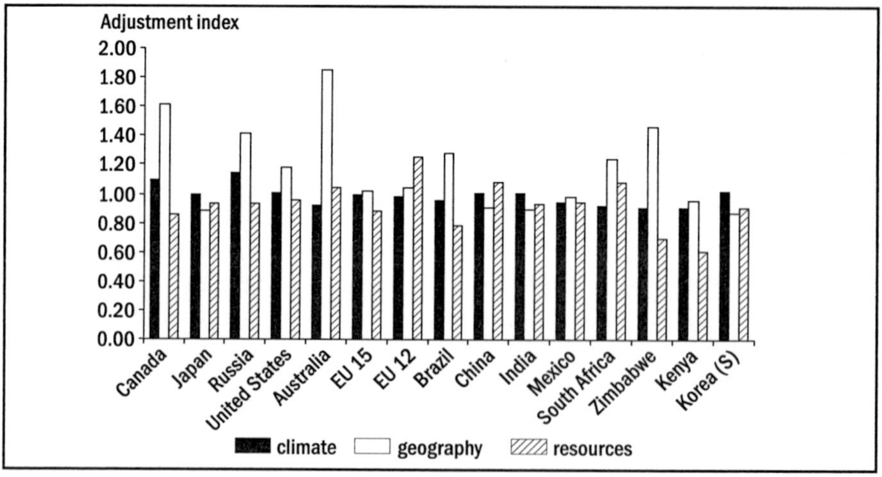

Figure 3 Adjustment coefficients for climate, geography and resources endowment

Source Authors' calculation based on data in Tables 3, 4, and 5

changed. The overall coefficient to adjust the initial global carbon budget through history to future is used for different emission scenarios. Figure 4 displays the comparison between with and without adjustment of carbon budget from year 1900 to year 2050 under Scenario A.

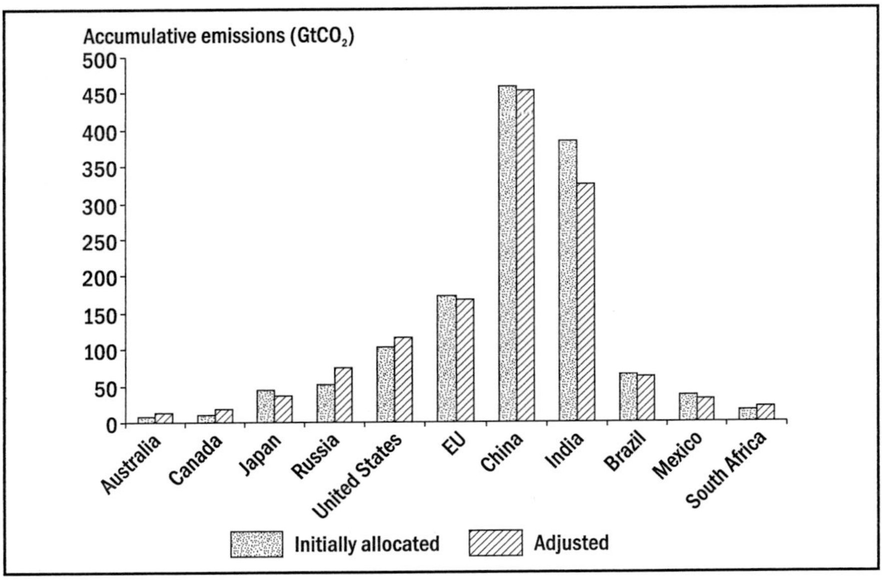

Figure 4 Overall adjustment of carbon budget

Note Gt – gigatonnes

Source Authors' calculation for adjustment based on data in tables 3, 4 and 5.

Arrangement of transfer payment

The adjusted carbon budgets in the aggregate for different countries are given in Table 6. Evidently, the accumulative historical emissions in developed countries represent a systematic overall deficit against their carbon budget substantially. Some Annex I countries have already used up their carbon budget for the future up to 2050. For instance, the accumulative historical emission in the US is about three times its total carbon budget. Countries with industrialization at an earlier stage such as the UK and Germany face similar problems, with considerable deficits as compared to their budget allocation.

Most developed countries have already overused their carbon budget and occupied others for their current consumption. Historical emissions occurred in the past and cannot be reversed. For one

Table 6 Transfer payment of carbon budget from the developing to developed countries

	Accumulated historical emissions ($GtCO_2$)	Natural factors adjusted carbon budget	Accumulated historical emissions/ total carbon budget	Balance of carbon budget ($GtCO_2$)	Carbon budget needed for future basic needs ($GtCO_2$)	Total transfer required for balance of carbon budget ($GtCO_2$)
Canada	23.0	17.2	2.03	(5.8)	5.2	11.0
France	28.1	17.2	1.31	(10.9)	5.2	16.1
Germany	72.0	26.2	2.48	(45.8)	8.0	53.8
Italy	17.2	18.0	0.83	0.9	5.5	4.6
Japan	44.7	36.2	0.99	(8.5)	11.0	19.5
Russia	90.0	76.0	1.79	(14.0)	23.1	37.1
UK	54.8	18.8	2.59	(36.0)	5.7	41.7
US	309.3	117.2	2.97	(192.1)	35.7	227.8
Australia	12.1	12.8	1.68	0.7	3.9	3.2
EU15	217.8	119.9	1.60	(97.9)	36.5	134.4
EU12	50.3	46.7	1.38	(3.7)	14.2	17.9
Annex I-39	788.1	478.1	1.78	(310.0)	145.6	455.7

Note Numbers in bold indicate historical emissions in excess of their total budget allocation; Gt – gigatonnes

Source Authors' calculation based on data in above tables.

thing, emissions in the past were not considered wrong doing due to limited knowledge of global warming. Furthermore, large amount of budget surplus in the developing countries in the past was not due to their awareness of environmental disasters. Therefore, it might be reasonable to use the surplus budget from the developing countries to cover the historical deficit by the developed countries. A few developing countries such as South Korea and Saudi Arabia have their accumulative historical emissions managed within their carbon budget but their remaining budget looks insufficient for them to meet their basic needs in the future.

While the first transfer payment from the developing countries is made to balance the historical deficit in the past, there is a need to consider a second transfer payment to cover basic needs satisfaction. Although there is no budget left in the developed countries for their future up to 2050, people in these countries have their right to emissions for their basic necessities. No matter whether one is in a rich or poor country, basic needs emissions must be considered an entitlement in line with the equity principle. As developing countries do have huge amount of budget surpluses, a second transfer payment can be made to support the basic needs by people in the developed regions where there is no budget for their emissions in the future.

Figures in Table 6 show that the accumulated historical emission from Annex I countries is 788.1 $GtCO_2$, 310.0 $GtCO_2$ in excess of their total budget, which should be 478.1 $GtCO_2$. The US and the EU are responsible for the largest shares, 62.0% and 32.8% respectively. Italy and Australia have a certain carbon budget left, but the remaining is not enough to meet their future basic need emissions. Assuming that carbon budget per capita over the period from 1900 to 2050 number at 2.33 tonnes of CO_2 represents the amount of emissions for basic necessities on world average, overall adjustment taking into account climatic, geographical, and resources endowment factors can arrive at a budget level for basic necessity emissions. In aggregate, the budget needed by Annex I countries to satisfy their basic need emissions are calculated at 145.6 $GtCO_2$. This amount of emissions budget has to be from the surplus budget in the developing countries.

Through coverage of historical deficit and basic necessity emissions for the future up to 2050, total volume of carbon budget transferred from Non-Annex I to Annex I countries would amount to 455.7 GtC_{O2}. Pricing at 10 euros/tonne CO_2, the current Certified Emission Reduction (CER) price in the international market, the value of budget transfer would reach 4.6 trillion, about 100 billion a year from now up to 2050, far beyond the amount of financial assistance from developed to developing countries. The main beneficiaries of the budget transfer payments are the US and the EU, accounting for 50% and 33.4% respectively.

As the amount of global carbon budget is fixed to ensure global sustainability, budget transfer to developed countries must come from surplus budget from the developing ones. Because of low level of economic development and limited historical emissions, most of the developing countries have large amount of budget surpluses (Table 7). Overall carbon budget surplus from Non-Annex I countries is estimated at 83%, and the surplus in India at 93% while that in Indonesia is 92%. Under the business as usual (BAU) scenario, the surplus could be transferred to the North. Assuming that emissions from Non-Annex I countries would peak in 2035, and increase by 60% as compared to the 2005 level, and then would drop gradually, with a final increase of 50% in 2050 over 2005 level. Based on this estimation, the accumulated emission of developing countries would be 1015.3 $GtCO_2$ with 462.8 $GtCO_2$ available for carbon budget transfer. Those countries with large population and relative low development level would contribute a lot to the carbon budget transfer. For instance, the natural factors adjusted carbon budget in India is at 324 $GtCO_2$ as compared to its accumulated historical emission 27.1 $GtCO_2$. According to projections by the IEA (2007), emissions in 2030 would be 76% higher than that in 2005 and 105% in 2050. This means that the accumulated emissions in the future would be 104.8 $GtCO_2$ with a budget surplus of 192.1 $GtCO_2$. And this part of carbon budget transferred to the developed countries would not bring negative effects to development in the South. Furthermore, the financial mechanism in the future would arrange funds distribution according to the volume of carbon budget transferred.

For some rapidly industrializing developing countries, the carbon budget would also constitute a constraint to their further development, because their accumulated emissions in the future are likely to exceed the carbon budget under the BAU scenario. Take China as an example. China's emission is estimated to peak in 2030, an increase by 105% over 2005 as the base year. Then it is projected to decline slowly. By 2050, emissions would be 90% of the level in 2005. With the accumulated emissions in future, 445.6 $GtCO_2$, and its accumulated historical emissions, 88.7 $GtCO_2$, China is likely to exceed its carbon budget at 454.2 $GtCO_2$. Countries with same situation would not contribute to the carbon budget transfer and a low-carbon development pathway should be taken to make sure that the accumulated emissions in the future could be able to meet the carbon budget target.

The change of carbon budget for Annex I countries with transfer payment from Non-Annex I countries is illustrated in Figure 5. Figure 6 shows annual carbon budget measured by per capita. Compared with the historical emissions per capita in 2005, which is 1.63 tonnes of CO_2, historical emissions in developed countries are far higher than the initial budget allocated. For instance, emissions per capita in the US are at 9.94 tonnes CO_2, 8.67 tonnes CO_2 in the UK and 8.32 tonnes CO_2 in Germany. After the transfer of carbon

Table 7 Potential contributions to carbon budget transfer by developing countries

	Accumulated historical emission	Natural factors adjusted carbon budget	Accumulated historical emission / total carbon budget	Balance of carbon budget	BAU* (%, compared to 2005)	Accumulated emission in future (BAU)	Potential carbon budget available for transfer
Non-Annex I	316.8	1794.9	0.17	1478.1	62% in 2035 peaking, 50% in 2050	1016.8	461.3
Brazil	8.5	62.1	0.14	53.6	57% in 2030 65% in 2050	22.0	31.6
India	27.1	324.0	0.07	296.9	76% in 2030 105% in 2050	104.8	192.1
Indonesia	6.3	66.1	0.08	59.7	70% in 2030 110% in 2050	31.1	28.6
Nigeria	2.2	29.1	0.04	27.0	47% in 2030 70% in 2050	7.4	19.6

Note The low-growth scenario in the IEA's *International Energy Outlook 2007* was taken for estimate of the numbers by 2030.

Source Authors' calculation based on data from IEA (2007) and the above tables.

budget from developing to developed countries, total carbon budget acquired by developed countries is still far in excess of the global average level. Taking the US as an example, carbon budget is increased to 7.71 tonnes of CO_2 per capita after transfer payment as compared to 2.33 tonnes of CO_2 per capita under initial budget allocation.

Figures in Table 8 indicate that due to the large population in developing countries, with 80% of the global population, the share of initial allocation of carbon budget between Annex I and Non-Annex I is 19.5% vs 80.5% The natural factors adjustment may have significant influence for certain countries, but such adjustment has limited impact on the overall pattern of budget allocation between Non-Annex I and

Annex I countries. But if transfer payment of carbon budget from Non-Annex I to Annex I is brought into account, the share of total budget allocation between Annex I and Non-Annex I comes to 40.5% vs 59.5%, with Annex I Parties being more than doubled.

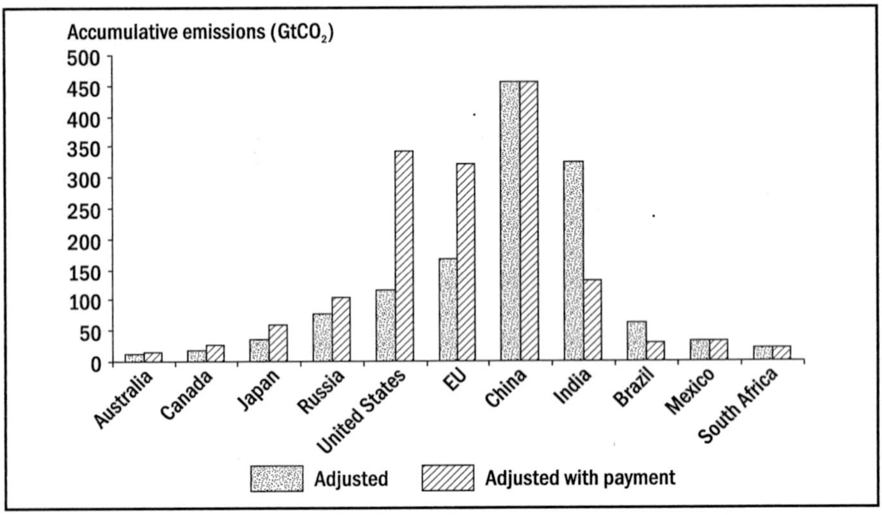

Figure 5 Comparison of adjusted carbon budget with or without transfer payments in selected countries (in accumulative emissions)
Source Authors' calculation

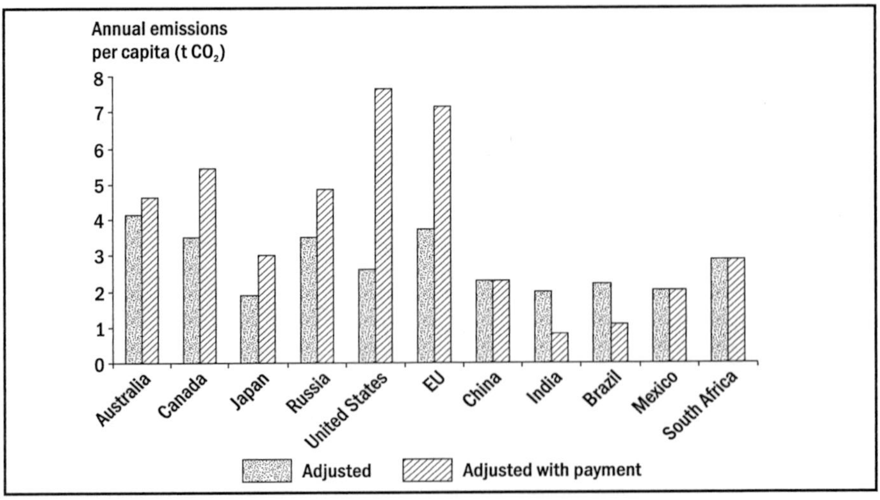

Figure 6 Comparison of adjusted carbon budget with or without transfer payments in selected countries (in average emissions per capita per year)
Source Authors' calculation

Table 8 Percentage shares of carbon budget between Annex I and Non-Annex I countries (1900-2050)

	Initial carbon budget	Natural factors adjustment	Adjustment with payment
Non-Annex	80.5%	79.0%	58.9%
Annex I	19.5%	21.0%	41.1%

International elements for implementation and compliance

Initial, natural factors adjusted and payment transferred carbon budget allocation for different countries could be decided following certain rules for calculation. The next part for building a global climate regime lies in the design of implementation mechanisms for operationalizing such a carbon budget framework, so as to stimulate countries for control of emissions in line with allocated carbon budget and to actively contribute to global actions for addressing climate change. For global climate protection, global long-term target would only be achieved by countries meeting their carbon budget. An effective international climate regime must include elements of market, financing, measurement, reporting and verification, compliance and the like.

Market mechanism

Market mechanism is an important part of an international climate regime. Though the calculation of carbon budget carefully considers realistic national circumstances for satisfaction of basic needs in different countries, actual emissions can highly deviate from the budget as allocated at certain stages of development. Therefore, an international carbon market is needed for the transaction of budget through trading for an efficient re-allocation of the budget.

As Annex I countries have accumulated emissions in the past, there is a huge budget deficit despite strict emissions reduction measures being adopted in the future. Consequently, they need to purchase emissions credits from international market since their carbon budget per capita is in excess of international average level. Table 9 indicates that their carbon budget after adjustment and transfer payment would be 923.1 $GtCO_2$. But their historical emission alone is at 788.1 $GtCO_2$ and the accumulation of emissions in the future would amount to 328.6 $GtCO_2$, assuming emission reduction to 40%

in 2020, 60% in 2030, and 85% in 2050 as compared to the 2005 level. The difference between demand and allocation gives a number of the potential demand for the carbon budget, which is estimated at 193.63 $GtCO_2$, about 58.9% of the future accumulated emissions. Given that Annex I countries could realize the emissions reduction target through emissions pathways as specified above, the US, Canada, and Australia would require to purchase 70% for their future emissions, while EU member states would have a lower demand in the carbon than their North American counterparts, about 44% on average for meeting their emissions requirements in the future. France requires less purchase of carbon credits in the market, at about 25.2% of its total future emissions. Global total carbon budget could be brought in balance if Annex I countries reduce their emissions targets as specified below (Table 9). But the demand in the international carbon market would be exceptionally high if emissions reduction target by Annex I countries could not be realized.

Table 9 Potential demands for carbon credits in Annex I countries

	Adjusted CB with payment (A)	Mitigation scenario (as compared to 2005 emissions)	Future accumulative emissions ($GtCO_2$)	Potential demands ($GtCO_2$)	Share (%)
Annex I-39	923.1	2020–40% 2030–60% 2040–85%	328.6	193.6	58.9
Australia	14.2		8.5	6.3	74.3
Canada	26.5		12.4	9.0	72.1
Japan	58.4		28.4	14.7	51.8
Russia	105.3		34.7	19.4	55.8
France	34.6		8.7	2.2	25.2
US	341.1		133.4	101.6	76.2
EU	320.7		93.1	40.7	44.0

Note Assuming that all these countries would reduce their emissions proportionally as compared to 2005 emission level, 40% in 2020, 60% in 2030, and 85% in 2050.
CB – carbon budget

Source Authors' calculation

Non-Annex I countries would be the main supplier of carbon credits in global carbon market. Two ways are available to acquire the saleable emission credits. One is domestic emissions reduction actions and the other is international cooperation. However, hot air created under carbon budget is not tradable as the budget is for development and not for cash profit though this part could be used for transfer payment of carbon budget. On a project-based manner, real reductions would be allowed for trading. The total carbon budget after overall adjustment for natural factors for Non-Annex I countries would be around 1794.9 $GtCO_2$. In order to make the transfer payment to Annex I countries, emissions in Non-Annex I countries should peak by 2035, with an increase of 60% over 2005 level, and then emissions are required to decrease gradually to 50% in 2050 as compared to 2005 level. If further carbon demand of Annex I countries need to be satisfied, another possible scenario is that emissions in Non-Annex I countries peak at 2025, with an increase by 40% over the 2005 level and afterwards, drop dramatically down to 50% in 2050. Those countries with their carbon budget higher than BAU emissions would have more potential for transfer payments and carbon credits.

Take India as an example, as illustrated in Figure 7. Natural factors adjusted carbon budget is 324 $GtCO_2$, with historical emissions at 27.1 $GtCO_2$. This small amount of accumulative emissions indicates a potential for increase of emissions at 22% per year up to 2050 without going over its budget. Evidently, increase at such a rate is not possible at all. It is assumed that under BAU conditions, an amount of 192.2 $GtCO_2$ surplus in hot air conditions would be used for transfer payment of carbon budget. The remaining amount of carbon budget would be at 104.8 $GtCO_2$ for practical implementation. Considering that India is to develop a low-carbon economy or even lower carbon pathway through international cooperation, it is unlikely that India would increase its emissions by 50% in 2030 over its 2005 level, and by 60% in 2050. Thus, the accumulated emissions of India in future can be projected at 93.5 $GtCO_2$. This would mean that India would generate 11.3 $GtCO_2$ for sale. But for countries with carbon budget below their BAU, the picture can be different. China, for instance, would face a carbon budget constraint for its future development, despite the fact that emissions reduction actions are taken to make emissions pathways deviated from BAU. As a result, China is likely to be a net importer of carbon credits instead of a net exporter under carbon budget constraint. Therefore, greater efforts will have to be made in order to reduce the risk of overshooting the budget constraint in China. The real scale of international carbon market will depend on the demand–supply relationship and emissions reduction efforts taken by countries. If demand is high and supply is insufficient, carbon price will increase, which would stimulate low-carbon investment for the generation of more carbon credits for sale. Such investment creates substantial opportunities for international cooperation through emissions trading with effective emissions reductions at lower cost under an overall carbon budget constraint.

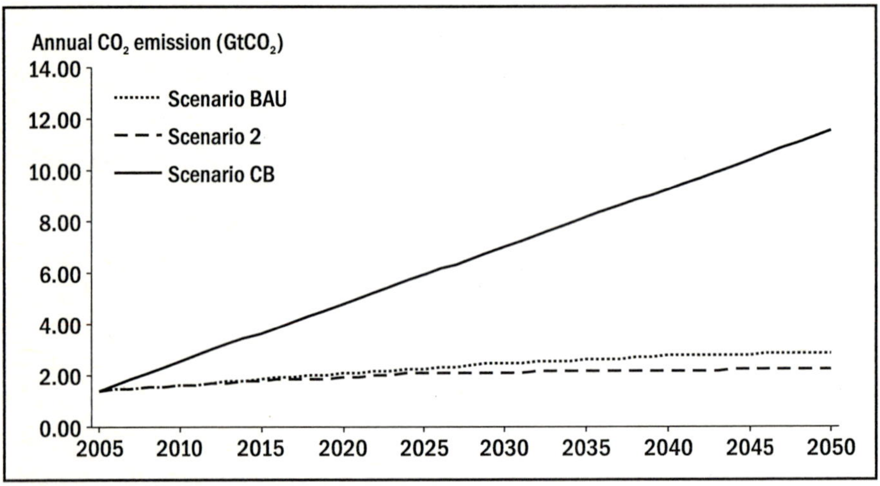

Figure 7 Carbon budget and emissions scenarios in India

Financial mechanism

If actual emissions by a country exceed the sum of carbon budget and carbon credits purchased from the international carbon market, a progressive carbon tax is recommended as a compliance mechanism with financial implications. The base for the progressive carbon tax would be conditional on the degree of excessive emissions over the carbon budget. The higher the incremental amount of emissions, the higher the rate of progressive carbon tax. However, the rate of the tax is capped by the price of renewable energy. If the rate is equal to the price of renewable energy, the buyer of carbon credits would go for renewable energy for substitution of traditional fossil fuel energy instead of paying for the penalty tax. The rate of progressive carbon tax at incremental level is illustrated in Table 10. For instance, carbon budget for the US is 31.8 $GtCO_2$ from now up to 2050. This means that the US would have to purchase an amount of 101.6 $GtCO_2$ from the international carbon market, assuming the US would meet emission reduction targets of 40%, 60%, and 85% in 2020, 2030, and 2050 respectively over its 2005 emission level. If the global carbon market is able to supply only half of its demand, the accumulative emissions in the US would be 82.6 $GtCO_2$, 2.6 times higher than its carbon budget. The data in Table 10 illustrates that the total amount of progressive carbon tax from 2005 to 2050 would be $200–400 billion, at $4.3–8.7 billion per year if CO_2 is priced at $5–$10 per tonne of CO_2.

Table 10 Progressive carbon tax

μ=E/CB		Tax rate (the price of renewable energy=λ)	Taxable revenue, taking the US as an example ($ billion)	
			λ=5	λ=10
0<μ≤1.0	Basic need satisfaction	0	0	0
1.0<μ≤1.5	Limited excessive emissions	0.5 λ	39.75	79.5
1.5<μ≤2.0	Moderate excessive emissions	0.8 λ	63.6	127.2
μ≤2.0	Severe excessive emissions	λ	95	190
Total			198.35	396.7

Source Authors' calculation.

Several ways could be used for a country to realize its carbon budget target. One is domestic emissions reduction. The second should be buying or selling emissions credits by taking part in international market or international cooperated emissions reduction actions. The third is to pay for the penalty progressive rate of tax. But to guarantee the realization of global environmental targets, emissions in excess of the budget after being penalized should be included in the calculation for the next commitment period. Government decision-makers should make the most reasonable choice and minimize the reduction cost after weighing the advantage and disadvantage in accordance with their national circumstances.

The tax revenue could be injected into current funding mechanisms or used to establish a new global fund in order to support mitigation and adaptation action taken by the developing countries and promote technological transfer. The contribution of countries to payment transfer of carbon budget should be taken into account when the fund is applied and allocated. Countries such as India, which avail more carbon budget for transfer payment should take priority allocation of the monetary transfer for mitigation actions.

Verification and compliance mechanisms

In general, the mechanism design under the CBP is compatible with the mechanisms set in the Kyoto Protocol.

Commitment periods

The mechanism is designed for implementation over the period from 2005 to 2050 divided by a few commitment periods according to the progress of negotiations. For instance, the second commitment period could be set from 2013 to 2020, taking 2005 as the base year.

Enlarging market mechanism

After separating hot air from carbon budget, all the countries could participate in the global carbon market.

Strengthening financial mechanism

The current financial mechanism is on a voluntary basis, but the scale of fund designed in the CBP is enlarged, and the mechanism is made compulsory.

MRV

As the process of carbon budget allocation, adjustment, and transfer is transparent and predictable, only regular evaluation is needed on the implementation of budget, using emissions data for reporting. No additional difficulties would be added to the measurable, reportable, verifiable (MRV) process.

Enhancing compliance mechanism

The current compliance mechanism is weak. It can be enhanced in the carbon budget proposal by introducing a compulsory penalty (progressive carbon tax) mechanism.

Implications for China

As initial allocation of carbon budget is based on the number of people, one may suggest that this proposal is in favour of the most populous countries in the world such as China and India. It does not have to be the case, however. As a matter of fact, the CBP imposes a tight emission constraint for China to meet its development target. This constraint is likely to provide pressure and incentives for China to follow a path of low-carbon development.

The carbon budget of China after adjustment for national circumstances amounts to 452.2 $GtCO_2$, which is slightly below the initial allocation of budget 458.8 $GtCO_2$. Overall, adjustment for three factors cancels one another and the balance of impact is negligible. Climatic and resource factors enlarge the budget a little bit while geographical elements make the budget slightly smaller. Up to now, China has emitted some 88.7 $GtCO_2$ into the air and the amount left for future budget is about 365.5 $GtCO_2$. As Figure 8 shows, if China's emission reaches 105% peaking in 2030, and increase 90% in 2050 as compared to the 2005 level, future emissions under BAU are projected to be at 445.6 $GtCO_2$. This means that China's carbon budget would be exceeded by about 80.1 $GtCO_2$ under Scenario 1. According to Scenario 2, if China's emission increases 55% in 2030 and 45% in 2050 over 2005 level by means of developing low-carbon measures and international cooperation, China could only barely manage to meet its carbon budget, with nothing for transfer payment or sales in the carbon market, unless further emission reduction is made.

Both scenarios 1 and 2 are the statistic estimates, and prediction is based on production. As a world workshop, China generates a large proportion of emissions within its political boundary for consumption in other countries through international trade, at 30% or so in net terms (Pan and Chen 2008). This is an indication that China's emission is actually overestimated about 30% if measured from consumption side. This mismatch of emissions may disappear by enhancing its export structure in the long term, but it is hard to change China's situation in the world market in the near future. As Figure 8 shows, if the budget is calculated from the consumption side, the future accumulative emissions by China may be reduced by about 8% as compared to the production side accounting. Considering that the gap between both sides exists before 2030, it is unlikely that China would gain extra benefit from such a proposal.

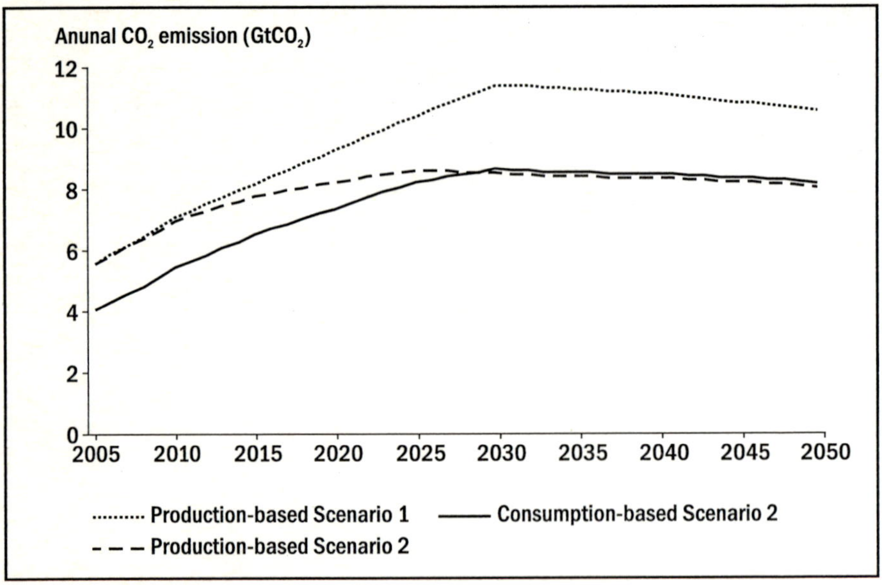

Figure 8 Emission scenarios for China based on production and consumption
Source Authors' calculation

Conclusions

In summary, carbon budget is defined as per capita cumulative emissions during a certain period of time for evaluation or an annual average during an evaluation period. Total global historical, current, and future cumulative carbon emissions are taken as the global budget in aggregate. This aggregate amount is the absolute budget available for the entire population on the earth to ensure climate protection. For equity considerations, this budget is allocated to individuals with adjustments taking into account natural factors such as climate, geography and resource endowment. This budget line of carbon emissions would ensure both environmental sustainability and inter-personal equity. In other words, climate goal should take priority over development targets so as to avoid an irreversible climate disaster one day.

This proposal is in contrast to many others in the existing literature. The budget is made for individuals although it can be aggregated to arrive at a national total. But its essence is not based on international equity as measured by political entities. Many key indicators such as GDP, energy consumption, emissions, and so on, are temporary factors. They change over time and should not be used as key determinants for allocation of emission entitlements. There is no need to go a long way for emission allocation to meet the demand by people. Therefore, carbon budget constitutes a solid foundation for sustainable and equitable emission allocation.

Simply speaking, the CBP establishes a threshold for per capita cumulative emissions which would meet a long-term goal and reflect the differential conditions between countries. Carbon budget represents that everyone should make efforts to keep his/her 'carbon footprint' in a reasonable range. Both the developed and developing countries have responsibilities to adopt appropriate policies and measures to encourage sustainable consumption through meeting basic needs for both the poor and the rich by discouraging emissions due to luxurious and wasteful life style. Given the fact that some people have already used up their budget while others have substantial surplus budget, payment transfer of carbon budget is possible to meet the basic needs for those with deficit and to generate financial resources for low-carbon development. Therefore, effective emissions trading would lead to win-win solutions in the fight against climate change.

The methodology of the CBP still needs to be improved. For instance, cumulative emissions in each case are simply piled up by each year. Scientifically, global warming potential of CO_2 once emitted diminishes over time and such a factor should be considered in the calculation. However, accurate calculation requires the support of climate modelling. Some uncertainty, however, would arise from data without actual observed values, especially in the future emissions scenarios. From a qualitative point of view, with a great amount of historical emissions (which means a higher level in stage of development), developed countries have more technological and financial advantages on mitigation than the developing countries. While historical accumulation of emissions in the developing countries is relatively small, they have to require more emissions in the future to meet their basic needs, in particular the amount of carbon embodied in physical build-ups such as infrastructure and buildings. Nevertheless, latecomers may have technological spillover advantages as the amount of carbon now for one tonne of steel produced is much less than that produced one hundred years earlier. These two factors deserve further examination for understanding the impact of budget allocation.

The selection of some parameters such as long-term target of global mitigation, base year for calculating historical emissions, BAU for developing countries, and so on, may meet challenges from a different perspective. While we acknowledge such controversies, what should be emphasized is the fact that indicative experiment is helpful to understand the process and impacts. To a certain extent, some of the parameters may be subject to political negotiation. But this does not mean that analytical work should await instructions from politicians or negotiators.

Analytical result in this report reveals the following facts for the global community's consideration. First, a global target for a 50% reduction by 2050 could be used as the absolute limit for global carbon budgeting. The developed countries have already used up their budget in the past, and in future, they have to rely on transfer payment of carbon budget from the surpluses by their developing counterparts. Second, developing countries in general have budget surpluses, which can be used as an asset for balancing emissions demand at global level. Such surplus budget can be transferred to the developed countries and in return, financial support can be availed for low-carbon development. Finally, for building an equitable and sustainable global climate regime, the basic elements of this carbon budget proposal can be a sound basis for meeting the long-term stabilization target and development needs through international cooperation.

Acknowledgements

The authors acknowledge with thanks the constructive comments and suggestions from Sivan Kumar of SEI, Kelly Gallaghe and Rob Stowe of Harvard Kenny School of Government, Robert Heilmayr of WRI and many others at our workshops in Beijing, Boston, and Washington DC. Wang Wenjun and Li Chenxi at CASS/RCSD contributed actively to the work with respect to data analysis and literature review. Our colleagues at CASS/RCSD, Chen Hongbo and Chen Bingbo, also contributed substantially in the process of the research. Zheng Yan and Liao Maolin helped with translation of part of the report. Nevertheless, the views and errors remain the responsibility of the authors. The authors also would like to express gratitude to the United Nations Development Programme (UNDP) and National Development and Reform Commission (NDRC), for their financial support for this project, which made it possible to carry on and complete this paper.

References

Aslam M. 2002
Equal per capita entitlements: a key to global participation on climate change?
In *Building a Climate of Trust: the Kyoto Protocol and beyond*, edited by K Baumert, O Blanchard, S Llosa, and J Perkaus
Washington, DC: World Resources Institute

Baer P and Mastrandrea M. 2006
High Stakes: designing emissions pathways to reduce the risk of dangerous climate change
London: Institute for Public Policy Research (IPPR)
[Details available at <http://www.ippr.org/members/download.asp?f=%2Fecomm%2Ffiles%2Fhigh%5Fstakes%2Epdfhttp://www.ippr.org>, last accessed on 18 September 2009]

Baer P, Athanasiou T, and Kartha S. 2007
The Greenhouse Development Rights Framework: rationales, mechanisms, and initial calculations
Berkeley: EcoEquity and Christian Aid
[Details available at <www.ecoequity.org/docs/TheGDRsFramework.pdf>, last accessed on 23 April 2007]

Baer P, Harte J, Haya B, Herzog A V, Holdren J, Hultman N E, Kammen D M, Norgaard R B, Raymond L. 2000
Equity and greenhouse gas responsibility
Science 289 (5488): 2287

Baumert K and Selman M. 2003
Data note: heating and cooling degree days
Details available at <http://cait.wri.org/downloads/DN-HCDD.pdf>, last accessed on 18 September 2009

Bodansky D, Chou S, and Jorge-Tresolini C. 2004
International Climate Efforts Beyond 2012
Arlington: Pew Center on Global Climate Change
[Details available at <www.pewclimate.org/international/ about>, last. accessed on 3 May 2008]

Brazil. 1997
Proposed elements of a protocol to the UNFCCC, presented by Brazil in response to the Berlin mandate, FCCC/AGBM/1997/MISC.1/Add.3. Bonn, UNFCCC
[Details available at <http://unfccc.int/cop4/resource/docs/1997/agbm/misc01a3.htm>, last accessed on 2 July 2009]

Burtraw D, Palmer K, and Kahn D. 2009
A symmetric safety valve.
Details available at <http://www.rff.org/rff/documents/RFF-DP-09-06.pdf>, last accessed on 18 September 2009
[RFF Discussion Paper 09-06]

CDIAC. 2008
Global, regional and national fossil fuel CO_2 emission
Details available at <http://cdiac.ornl.gov/trends/emis/meth_reg.htm>, last accessed on 27 August 2008

Chen Y and Pan J. 2004
Equity concerns over climate change mitigation
Chinese Journal of Population Resources and Environment, **2** (2): 334–337

Chung R. 2007
A CER discounting scheme could save climate change regime after 2012
Climate Policy 7 (2): 171–176
Details available at <http://2050.nies.go.jp/3rdLCSWS/related/related_RaeKwonChung.pdf>, <http://www.earthscanjournals.com/cp/007/02/default.htm>, last accessed on 18 November 2009

Den Elzen M, Höhne N, and Moltmann S. 2008
The triptych approach revisited: a staged sectoral approach for climate mitigation
Climate Policy **36** (3): 1107–1124

Den Elzen M, Hoehne N, Brouns B, Winkler H, Ott H E, et al. 2007
Differentiation of countries' future commitments in a post-2012 climate regime: an assessment of the South–North dialogue proposal
Environmental Science and Policy **10**: 185–203

Edmonds J. 1998
The economics of climate change: building backstop technologies and policies to implement the Framework Convention on Climate Change
Energy and Environment 9 (4):383–397

Ellis J and Baron R. 2005
Sectoral crediting mechanisms: an initial assessment of electricity and aluminium
Paris: International Energy Agency/ Organisation for Economic Co-operation and Development
[Detail available at <http://www.iea.org/papers/2005/rb_sectoral.pdf>, last accessed on 18 September 2009]

Groenenberg H, Phylipsen D, and Blok K. 2001
Differentiating the burden world-wide: global burden differentiation of GHG emissions reductions based on the Triptych approach— a preliminary assessment
Energy Policy **29**: 1007–1030

Herzog T, Baumert K, and Pershing J. 2006
Target – intensity: an analysis of greenhouse gas intensity targets
Washington, DC: World Resources Institute
[Details available at <www.wri.org/publication/ target-intensity> last accessed on 8 May 2008]

Höhne N and Lahme E. 2005
Types of future commitments under the UNFCCC and the Kyoto Protocol post-2012 [Briefing paper for WWF]
Gland: Switzerland, Worldwide Fund for Nature
[Details available at <http://assets.panda.org/downloads/ecofyspost2012targets20sept05.pdf>, last accessed on 18 September 2009]

Höhne N, Den Elzen M, and Weiss M. 2006
Common but differentiated commitments (CDC): a new conceptual approach to long-term climate policy
Climate Policy **6** (2): 181–200

IPCC (Intergovernmental Panel on Climate Change). 2007
Climate change 2007: mitigation of climate change
[Contribution of Working Group III to the Fourth Assessment Report of the Intergovernmental Panel on Climate Change.]

IEA (International Energy Agency). 2007
World International Energy Outlook 2007
Paris: IEA
[Details available at <http://www.iea.org/textbase/nppdf/free/2007/weo_2007.pdf>, last accessed on 18 September 2009]

Kartha S, Athanasiou T, Paul Baer T, Kemp-Benedict E. 2008
A call for leadership: a greenhouse development right's analysis of the EU's proposed 2020 targets.
Details available at <http://www.indiaenvironmentportal.org.in/files/EcoEquity-2008-Repo.pdf>, last accessed on 18 September 2009

Kim Y G and Baumert K A. 2002.
Reducing uncertainty through dual-intensity targets
In *Building a Climate of Trust: the Kyoto Protocol and beyond*, pp. 109–34, edited by K Baumert, O Blanchard, S Llosa, and J Perkaus.
Washington, DC: World Resources Institute

La Rovere E, de Macedo L, and Baumert K, *et al*. 2002
The Brazilian proposal on relative responsibility for global warming
In *Building on the Kyoto Protocol: options for protecting the climate*, pp. 157–74, edited by K Baumert, O Blanchard, S Llosa, and J F Perkaus
Washington, DC: World Resources Institute

Meyer A. 2000
Contraction and Convergence: the global solution to climate change
Bristol: Green Books and Schumacher Society

Meyer A. 2004
GCI briefing: contraction and convergence
Proceedings of the ICE - Engineering Sustainability **157** (4)
[Details available at <http://www.gci.org.uk/briefings/ICE.pdf>, last accessed on 18 September 2009]

Pan J. 2002.
Understanding human development potentials and demands for greenhouse gas emissions: with empirical analysis using time series and cross-sectional data.
Social Sciences in China (in Chinese) (**6**)

Pan J. 2008
Carbon budget to meet basic needs and its implications on international equity and sustainability
World Economics and Politics (**1**) (in Chinese)

Pan J, Phillips J, and Ying Chen Y. 2008
China's balance of emissions embodied in trade: approaches to measurement and allocating international responsibility
Oxford Review of Economic Policy **24**: 354–376

Pan J, Xianli Zhu, and Ying Chen Y. 2005
Fulfilling basic development needs with low emissions—China's challenges and opportunities for building a post-2012 climate regime
In *Governing Climate: the struggle for a global framework beyond Kyoto*, edited by T T Sugiyama, published by IISD, December 2005

Pinguelli Rosa P and Kahn Ribeiro K. 2001
The present, past, and future contributions to global warming of CO_2 emissions from fuels: a key for negotiation in the climate convention
Climatic Change **48**: 289--308

Samaniego J and Figueres C. 2002
Evolving to a sector-based clean development mechanism
In *Building on the Kyoto Protocol: options for protecting the climate*, pp. 89–108, edited by K Baumert, O Blanchard, S Llosa and J F Perkaus
Washington, DC: World Resources Institute

Schmidt J, Helme N, Lee J, Houdashelt M, *et al.* 2006
Sector-based approach to the post-2012 climate change policy architecture
[Future Actions Dialogue Working Paper]
Washington, DC: Center for Clean Air Policy
[Details available at <http://www.ccap.org/docs/resources/539/CPOL8-5_05_Schmidt%20(2).pdf>, last accessed on 18 September 2009]

Sterk W and Wittneben B. 2006
Enhancing the Clean Development Mechanism through sectoral approaches: definitions, applications and ways forward
International Environmental Agreements: Politics, Law and Economics **6**: 271–287

Stern N. 2006
The Economics of Climate Change: the Stern Review
Cambridge, UK: Cambridge University Press

Stern N. 2008
Key Elements of a Global Deal on Climate Change
London: London School of Economics and Political Science
[Details available at <http://www.lse.ac.uk/collections/granthamInstitute/publications/KeyElementsOfAGlobalDeal_30Apr08.pdf>, last accessed on 2 July 2009]

Tangen K and Hasselknippe H. 2004
Converging markets
[Paper under the FNI/CRIEPI/HWWA/CASS post-2012 policy scenarios project]
Polhogda: Fridtjof Nansen Institute

UNFCCC (United Nations Framework Convention on Climate Change). 2002
Scientific and methodological assessment of contributions to climate change
[Report of the expert meeting. Note by the secretariat. FCCC/SBSTA/2002/INF.14] (related to proposal by Brazil). New Delhi.

Victor D G, House J, and Joy S. 2005
A Madisonian approach to climate policy
Science **309**: 1820–1821

Ward M. 2008
Sectoral approach(es): Ideas and 'agendas' [Presented at IPS Post 2012 Roundtable, 29 April, 2008]
Details available at <http://ips.ac.nz/events/completed-activities/Post%202012/Murray%20Ward%202008.pdf>, last accessed on 18 September 2009

Winkler H. 2008
Climate change mitigation negotiations, with an emphasis on options for developing countries. Capacity building for policy makers [UNDP project paper]. An Environment and Energy Group Publication
[Details available at <http://www.undp.org/climatechange/docs/English/UNDP_Mitigation_final.pdf>, last accessed on 18 September 2009]

Winkler H, Howells M, and Baumert K. 2007
Sustainable development policies and measures: institutional issues and electrical efficiency in South Africa
Climate Policy 7(3): 212–229

Winkler H, Spalding-Fecher R, Mwakasonda S, and Davidson O. 2002
Sustainable development policies and measures: starting from development to tackle climate change

WRI (World Resources Institute). 2003
Carbon Analysis Indicators Tool (CAIT)
Washington, DC: WRI
[Details available at <http://cait.wri.org>., last accessed in July 2008]

3

Challenges in climate change adaptation

Jessica Ayers and Saleemul Huq*

Anthropogenic climate change is a reality in present times (IPCC 2007). The impacts are beginning to be felt, and that those worst affected are and will continue to be the poorest people, in poor countries (Huq and Ayers 2007). One way of limiting the impacts of climate change is to find new (and foster existing) ways of coping with and adapting to the changing climatic conditions. Adaptation to climate change has, therefore, emerged as a key policy response, describing the adjustment in natural or human systems in response to actual or expected climatic stimuli or their effects, which moderates harm or exploits beneficial opportunities (IPCC 2007). Adaptation can be a process, action or outcome in a system (ecosystem, household, community, group, sector, region, country) to enable the system to cope with, manage or adjust to changing conditions, stresses, hazards, risks or opportunities, associated with climate change better (Smit and Wandel 2006). Adaptation is seen as one of the two options for managing climate change, the other being mitigation, which involves the limiting of greenhouse gasses (GHGs), particularly carbon dioxide and methane, to mitigate against further global warming. Adaptation and mitigation are inherently linked in the climate system: the more effective mitigation undertaken now, the less need for adaptation in the future.

* International Institute of Environment and Development (IIED), London

Until recently, international climate change efforts under the United Nations Framework Convention on Climate Change (UNFCCC) have largely concentrated resources and attention on mitigation. However, the mounting evidence for the implications of climate change on development, particularly in developing countries, coupled with the failures of industrialized countries to commit to and meet targets for adequate GHG reductions under the Kyoto Protocol, have given rise to the consensus that action on adaptation to climate change is now equally urgent. Since the inception of the UNFCCC in 1992, adaptation has gone from being the marginal policy option relative to mitigation, to being placed on equal footing as one of the four 'building blocks' of the future climate change regime to come out of the most recent climate change negotiations alongside mitigation, technology cooperation and finance (Ayers and Huq 2008a).

Despite this progress on adaptation, however, a number of challenges remain for both international and domestic climate change and development policy in building resilience to climate change. First, there is confusion over exactly what adaptation is, and is not, and how to (whether to) distinguish between vulnerability to the impacts of climate change, and vulnerability to climatic variability, and by extension how to discern climate change adaptation from 'baseline' development needs. Second, international funding for adaptation under climate change frameworks falls very short of projected costs of adaptation, and debates persist over alternative sources of funding and the role and responsibilities of development institutions in meeting the additional costs of adaptation (Ayers and Huq 2008b). Thirdly, there is a lack of clarity over how to do adaptation; how international funding for adaptation should be channelled to those most vulnerable, which institutions and practitioners are best placed to implement adaptation action, and how.

This chapter will consider each of these challenges in turn, and show that while action on adaptation is of course urgent, it is important that these challenges are taken up and considered comprehensively, to avoid the misuse or duplication of scarce resources for adaptation, with potentially maladaptive results. First, we will demonstrate the need for effective action on adaptation by outlining the impacts of climate change, focusing on the most vulnerable developing countries.

Why is adaptation necessary?

In the recent *Fourth Assessment Report* of the Intergovernmental Panel on Climate Change (IPCC), observational evidence is used for the first time in the history of the IPCC to show that natural and human systems are already being affected by climate change (IPCC 2007). This means that, regardless of measures to cut GHG emissions today, the impacts of climate change are already inevitable as a result of past emissions that continue to have an impact today. Although it is too early to identify established trends in these observed impacts, the IPCC notes that early effects of climate changes and climatic variations on human and natural systems already include, among others, settlements in mountain regions being at enhanced risk of glacier lake outburst floods caused by mountain glaciers, and warmer and drier conditions in Sahelian regions of Africa resulting in reduced length of the growing season having a detrimental effect on agricultural production.

While climate change presents global risks, these risks are not distributed evenly (Dodman, Ayers, and Huq 2009). As noted by the IPCC, 'the distribution of impacts and vulnerabilities is still considered to be uneven, and low-latitude, less-developed areas are generally at greatest risk due to both higher sensitivity and lower adaptive capacity' (Scheider, Semenov, Patwardhan, *et al.* 2007, p. 781). The heaviest impacts will be felt in the least developed countries (LDCs), the small island developing states (SIDS) and Africa, which taken together form one group of 100 nations, collectively housing more than a billion people, but with their carbon dioxide (CO_2) emissions (excluding South Africa's) accounting for only 3.2% of the global total (Huq and Ayers 2007). This reflects profound global inequalities: that those countries that have contributed least to the problem of climate change will be worst affected by it, while those who have profited from high levels of GHG emissions are least threatened by the consequences of their actions (Dodman, Ayers, and Huq 2009).

The reasons for the uneven distribution of climate change risks are two-fold. First, because climate change will exacerbate the intensity and frequency of existing 'natural disasters', those countries that are physically vulnerable to natural disasters will likewise be the most physically vulnerable to climate change. Climate change vulnerability mirrors the existing uneven distribution of disaster risk, for example, in 2007, 37% of the reported disasters and 90% of all the reported victims occurred in Asia (Scheuren, le Polain, Below, *et al.* 2008).

The second and more poignant reason is that the ability of a country to cope with the impacts of climate change depends heavily on the development status of that country and the communities within it. For example, the Netherlands and Bangladesh have similar physical vulnerability contexts to climate change, both being flat low-lying countries threatened by sea level rise. Yet, the Netherlands has an extensive and well-developed dyke network to protect its coastal shoreline, while Bangladesh, being one of the LDCs, does not have the resources or capacity to build adequate sea defences to protect its population in the same way (Box 1). Within countries, extreme inequalities can serve to perpetuate vulnerability among lower income groups (LIGs), as the poor communities of New Orleans who experienced Hurricane Katrina in August 2006 can testify.

Vulnerability to climate change, therefore, depends on factors linked to development such as access to economic, ecological, social and human resources, and adequate institutions, governance, and infrastructure (Ayers and Huq 2008b; Huq, Reid, and Murray 2006; Klein, Schipper, and Dessai 2003; Klein, Huq, Denton, *et al.* 2007). As with all 'natural' hazards, people are only vulnerable when they are unable to cope with the impacts of hazards, depending on the extent to which they are socially, economically, and physically vulnerable. Based on this assertion, in its *Third Assessment Report*, the IPCC outlines the prerequisites for the capacity to cope with and adapt to climate change, to include a stable and prosperous economy, a high degree of access to technology at all levels, well delineated roles and responsibilities for implementation of adaptation strategies, systems in place for the national, regional and local dissemination of climate change and adaptation information, and an equitable distribution of access to resources (IPCC 2001). This presents a significant challenge for adaptation practitioners: that those countries most vulnerable to climate change also have the least capacity to adapt to it.

This challenge is compounded by the fact that the impacts of climate change threaten development, which in turn further undermines adaptive capacity. For example, if we consider the Millennium Development Goals (MDGs), eradication of extreme poverty and hunger is threatened by climate changes related to droughts and floods which would impact severely on agricultural crop production, reducing

Box 1 Climate change impacts in Bangladesh

Many of the projected impacts of climate change will reinforce the baseline environmental, socio-economic and demographic stresses already faced by Bangladesh. Climate change is likely to result in:

- Increased flooding, both in terms of extent and frequency, associated with sea level rise, greater monsoon precipitation and increased glacial melt.
- Increased vulnerability to cyclone and storm surges.
- Increased moisture stress during dry periods leading to increased drought.
- Increased salinity intrusion.
- Greater temperature extremes

The impacts of climate change on Bangladesh have significant implications for its development. The sectoral impacts of climate change include the following.

Agriculture

The economy of Bangladesh is based on agriculture, with two-thirds of the population engaged in or indirectly relying on agricultural activities. On average, during the period 1962–88, Bangladesh lost about 0.5 million tonnes of rice annually as a result of floods. Rice accounts for nearly 30% of the country's average annual food grain imports. Other impacts of climate change such as temperature extremes, drought, and salinity intrusion, are also causing declining crop yields in Bangladesh. The shortening of the winter season is resulting in a decline in production of winter crops. Salinity intrusion experienced by the coastal area of Bangladesh is having serious implications for the quality of the soil in areas that were traditionally used for growing rice. Under a moderate climate scenario the decline in yields due to salinity intrusion could be 0.2 million tonnes, which increases to 0.56 million tonnes under more severe scenarios. The IPCC FAR note that the production of rice and wheat could fall by 8% and 32% respectively by 2050 as a result of water stress.

Coastal areas

Most of Bangladesh is less than 10m above sea level, with almost 10% of the country below 1m, making it extremely vulnerable to increasing high tides. With sea levels expected to rise by an average of 2–3 mm per year during the first part of this century, the effects on the coastal areas will be severe, and include erosion, coastal land subsistence, siltation of river estuaries, reduced sedimentation, waterlogging, and saltwater intrusion. Cyclonic storms are also

Continued...

Box 1 Climate change impacts in Bangladesh *(Continued...)*

set to be exacerbated by climate change. Although Bangladesh now has good early warning systems and cyclone shelters have been constructed along much of the coast, infrastructure and livelihoods are still threatened and severely affected, hampering further development of the coastal areas. Thirty districts were damaged by Cyclone Sidr, for example, with the 11 districts closest to the coast damaged most severely.

Human health

The most direct impacts of climate change on human health occur through extreme events, for example, the floods in Bangladesh in 2004 caused 800 deaths, while Cyclone Sidr affected more than 8.5 million people, causing more than 3500 deaths. Climate change will also affect the distribution of climate-sensitive diseases. Incidences of malaria have increased dramatically in Bangladesh over the last 30 years, and other diseases such as dysentery, diarrhoea, dengue, hypertension associated with heat stress, asthma and skin diseases are also increasing in Bangladesh, particularly during the summer months. For example, increased flooding, as well as drought, is resulting in a decline in the availability of clean water, for a country where water-borne diseases are already responsible for 24% of all deaths.

Urban areas

Cyclone Sidr affected the infrastructure of more than half a million homes, with nearly one million all or particularly destroyed, and more than 10,000 schools all or partially destroyed. Severe flooding has already impeded the development of Dhaka significantly, but of the eight major floods that have occurred in the last 50 years, the three most recent (1988, 1998, and 2004) have been the most damaging. The key sectors affected by floods in Bangladesh's cities include infrastructure, industry, trade, commerce and utility services, all of which reduce in productivity during and after major flooding, increasing the vulnerability of the urban poor. Further, as the adverse impacts of climate change on rural areas cause increased migration to urban areas in search of non-agricultural employment, they also put greater pressure on scarce housing, water, sanitation, and energy services and increase the number of vulnerable urban poor who are particularly at risk from climate-related disasters. Already, around 40% of the population in Bangladesh live in slums and squatter settlements in the cities, which are most at risk from damage during flooding.

Source Adapted from Huq and Ayers (2008a)

the accessibility of food in the poorest societies; also, combating HIV/AIDS, malaria and other diseases is threatened by the impacts of climate change on access to clean water and the onset of new patterns of diseases induced by a changing climate (Levina 2007). Climate change also poses a threat to the long-term sustainability and efficacy of current development investments, and the World Bank estimates that up to 40% of development financed by overseas assistance and concessional loans is sensitive to climate risk (Burton, Diringer and Smith 2006).

Facilitating adaptation to climate change is, therefore, imperative in vulnerable developing countries to ensure that the development process is not further impeded, and vulnerability exacerbated. The priority for adaptation practitioners is building the resilience of the poorest and most vulnerable groups, and requires addressing both climate change and development elements of adaptive capacity. Development trajectories under a changing climate must be climate resilient, as climate change exacerbates old and brings about new challenges for development. It is vital that adaptation assistance is used to break the vulnerability-poverty nexus. Given that those who are most vulnerable to climate change are also least responsible for causing it, the responsibility for assisting adaptation in developing countries lies with developed industrialized nations who bear the greatest historic liability for the global GHG burden. This challenge of supporting poor countries in adapting to climate change must be taken up by high-income nations, based on the 'Polluter Pays Principle', pointing towards responsibility-based rather than burden-based criteria (Ayers 2009).

The importance of the adaptation agenda globally, and for the most vulnerable developing countries specifically, is clear: without adaptation, business-as-usual development is simply not sustainable under a changing climate. That the responsibility for supporting the adaptation burden lies with the industrialized nations, is also unquestionable. However, it is equally important that the urgency of adaptation does not overshadow the need to carefully consider how vulnerable developing countries can best be helped to adapt: understanding what sort of assistance is required, how much, and how this should be implemented is also a significant challenge.

Understanding adaptation

Meeting adaptation needs first requires understanding what those needs are. Yet, there is remarkable inconsistency when it comes to understanding adaptation, presenting a real challenge to adaptation policy and practice. As stated, adaptation has been defined by the IPCC as any adjustment in natural or human systems in response to actual or expected climatic stimuli or their effects, which moderates harm or exploits beneficial opportunities (IPCC 2007). Interpretations of this definition vary hugely however, and include policy-based institutional arrangements, for example, agricultural and land tenure policies; public/private investment in technologies such as building infrastructure, irrigation systems or large-scale embankments; and more recent livelihood-based approaches, which overlap (and indeed many argue are interchangeable) with development activities such as community-based low-tech irrigation systems, through to decreasing social vulnerability, for example, with livelihoods diversification programmes. The IPCC presents a wide variety of adaptation options, 'ranging from purely technological (such as sea defences), through behavioural (such as altered food and recreational choices), to managerial (such as altered farm practices) and to policy (such as planning regulations)' (IPCC 2007). Smit, Burton, Klein, *et al.* (2000) suggest that while there is general agreement that adaptation interpretations include 'adjustments in a system in response to... climatic stimuli, but they also indicate differences in scope, application and interpretation of the term adaptation' (2000: 228). Variations include 'adaptation to what?' Which can refer to simply climate, climate variability, or climate change; and 'Who or what adapts?' Which might be people, social or economic sectors, processes, or system structures?

Within the international climate change frameworks of the UNFCCC, 'adaptation to what?' is interpreted as adaptation to climate change specifically (rather than to broader climate variability including climate change, or taking into account underlying 'drivers' of vulnerability stemming from development needs). This has given rise to a policy context that is scientific, technical, and environmental, and a 'technology based' view of adaptation has resulted in priority being placed on adaptation measures such as dams, early-warning systems, seeds and irrigation schemes based on specific knowledge of future climate change conditions (Klein 2008). Correspondingly, the UNFCCC treats adaptation in the narrowest sense, as an issue of climate change, with adaptation actions limited to changes that are proven to be anthropogenic and distinct from climatic variability (Ayers and Dodman 2009).

However, this interpretation of adaptation limits the extent to which it can contribute to broader and sustainable vulnerability reduction. This matters because, as already discussed, vulnerability to climate change depends significantly on the development context. Technology-based measures can only be partially effective if they do not also address non-climatic factors that are the underlying 'drivers' of vulnerability. Klein (2008) provides the example of improving a water-supply system where climate change is associated with increased drought, which can only be effective insofar as everyone has equal access to that system; if the unequal distribution of water rights or the price of water excludes certain users from the system, people will remain vulnerable to drought. The definition of adaptation as an issue of climate change has created policy frameworks that do not fit with defining adaptation in terms of sustainable development (Ayers and Dodman 2009).

Building adaptive capacity must, therefore, be based on the principals of sound development. In the context of the MDGs, reducing poverty, providing general education and health services, improving living conditions in urban settlements, and providing access to financial markets and technologies, will all improve the livelihoods of vulnerable individuals, households and communities, and therefore, increase their ability to engage in adaptive action (Levina 2007). Adaptation activities are, therefore, often regarded as synonymous with development activities, and key to good development practice (Huq and Ayers 2008a). As stated by Huq and Ayers (2008a), 'Good (or sustainable) development (policies and practice) can (and often does) lead to building adaptive capacity. Doing adaptation to climate change often also means doing good (or sustainable) development.'

Recognizing and incorporating this range of adaptation options, from the 'development based' to the 'climate change' based approaches, presents a significant challenge to adaptation practitioners. A recent report provides a useful framework for understanding the variety and magnitude of adaptation options and practices in relation to development (McGray, Hammill, and Bradley 2007). The report suggests that the range of adaptation activities described above may be framed as a continuum from 'development orientated' to 'climate change orientated'. At one end, the most vulnerability-orientated adaptation efforts overlap almost completely with traditional development practice, where activities take very little account of specific climate change impacts and instead increase general resilience. At the opposite end, highly specialized activities exclusively target distinct climate change impacts, for example, where technology-based UNFCCC interpretations of adaptation would lie (McGray, Hammill and Bradley 2007).

It is important that attention is given to integrating this range of interpretations of adaptation into policy in both climate change and development arenas. While the recent widespread recognition of the implications of climate change for developing countries have resulted in understandable urgency to define adaptation and implement actions to contribute to resilience, it is vital that good development underpins all adaptation practices, to avoid dysfunctional adaptation that excludes rather than incorporates fundamental underlying development objectives (Ayers and Dodman 2009).

Raising funds for adaptation

Estimating costs of adaptation

Financing adaptation presents two challenges to adaptation practitioners: first, understanding 'how much' funding is needed; and second, how to raise the funds to meet these costs.

Quantifying the costs of adaptation is very difficult, though not the least because of the confusion outlined above over what exactly constitutes adaptation, and therefore, what should and should not be included in cost estimates. For example, the UNFCCC's emphasis on defining adaptation to climate change as distinct from climatic variability requires trying to distinguish the additional costs that climate change will put on development, regardless of the baseline development needs in the first place. At the level of the climate negotiations, the distinction is important, because such information informs political questions surrounding costs and burden sharing (Huq and Ayers 2008a; Ayers and Dodman 2009). As noted earlier, funding for adaptation is the responsibility of high-income, high-emitting countries, to be paid to countries most vulnerable to the impacts of those emissions. This distinction between adaptation and development is, therefore, supported by many developing countries in order to prevent industrialized countries incorporating adaptation funding into development assistance and thereby avoiding providing new and additional funding for adaptation under the UNFCCC (Ayers 2009).

On the ground, however, distinguishing between the costs of adaptation to climate change versus adaptation to climatic variability more generally is difficult to achieve. How do you 'measure' the percentage of extreme weather events that may be induced by climate change as opposed to those that would have occurred anyway? What proportion of development investment that is resilient to the long-term implications of climate change 'count' as adaptation? Further, given that those countries most vulnerable to climate change are also amongst the poorest, many countries cannot afford to meet the basic development needs, so the offer of funding the 'additional' cost of adaptation or 'climate-proofing' of development becomes futile. To give a recent example, one of the projects identified by the National Adaptation Programme of Action (NAPA) of Tuvalu is coastal infrastructure to protect the shoreline from erosion, a problem regardless of climate change (and so an existing development need), but one exacerbated by climate change (so also an additional cost as defined by the UNFCCC). The NAPA project team had extreme difficulties calculating the 'adaptation' component of the infrastructure needs. Further, being a poor country, the 'baseline' infrastructure is not yet in place and funding is not available to pay for it. Thus, the offer to fund, as it were, the 'top section' of the infrastructure required to respond to 'additional' impacts of climate change, is absurd in light of the fact that co-financing to pay for the lower section cannot be found. The project is currently in limbo while co-financing is sought (Ayers and Huq 2008b).

Estimating the costs of climate change impacts and adaptation is, therefore, inherently problematic, reflected by the lack of quantified data on the topic (Agrawala and Frankhauser 2008), and the variety in the scale of estimates (sometimes in orders of magnitude) that have been undertaken. Estimates that do exist tend to have been based on strong assumptions, such as perfect foresight, and there are very few cross-sector studies that look at the cumulative effects within countries or the wider macro-economic consequences of impacts or adaptation (Klein and Persson 2008).

However, a number of organizations have attempted to calculate the costs of adaptation in developing countries. The World Bank estimates that climate proofing development investments (including Official Development Assistance and concessional finance, foreign direct investment and gross domestic investment) in developing countries alone will cost between $10 billion and $40 billion annually (World Bank 2006). However, this figure has been heavily criticized by non-governmental organizations (NGOs) and others for not taking into account the costs of climate-proofing existing supplies of natural and physical capital where no new investment is planned, the costs of financing new investments specifically to deal with climate change, or the costs to households and communities to fund their own adaptation needs (Action Aid 2007; Oxfam International 2007). More recent estimates by Oxfam that do acknowledge these factors put the costs of adaptation closer to $ 50 billion annually (Oxfam International 2007) while the United Nations Development Programme (UNDP) suggests this could spiral to $86–109 billion annually by 2015 (UNDP 2007). The longer it takes to implement an effective international agreement to halt and then reduce GHG emissions, the higher these costs of adaptation will be and the more likely that the limits to adaptation will be reached and exceeded (Ayers 2009).

What is clear, even from these imperfect estimates, is that the costs of adapting to climate change in the most vulnerable developing countries are in the tens of billions of dollars per annum, well in excess of development budgets of these countries and certainly greater than current levels of development assistance, presenting a major challenge to meeting adaptation needs.

Meeting the costs of adaptation

To meet the costs of adaptation, the main source of international funding for adaptation is through the UNFCCC. There are four funds relevant to adaptation under the UNFCCC: The Least Developed Countries Fund (LDCF) established under the UNFCCC to help developing countries prepare and implement their NAPAs; the Special Climate Change Fund (SCCF), also under the UNFCCC to support a number of climate change activities such as mitigation and technology transfer, but prioritizes adaptation; the Global Environment Facility (GEF) Trust Fund's Strategic Priority for Adaptation (SPA), which pilots 'operational' approaches to adaptation; and the Adaptation

Fund. The first three funds are under the UNFCCC and managed by the UNFCCC financial mechanism, the GEF (Global Environment Facility), while the Adaptation Fund has an independent governing board and sits under the Kyoto Protocol.

However, these funds are not adequate to meet adaptation needs in developing countries (Ayers and Huq 2008b, Möhner and Klein 2007). First, they simply do not amount to enough. As of March 2008, the total resources pledged to the LDCF, the SCCF, and the SPA totalled \$298.4 million (\$172.8 million to the LDCF, \$75.6 million to the SCCF, and \$50 million to the SPA) (GEF 2008). Donors have since delayed on these pledged commitments because of an alleged lack of adequate and accountable mechanisms in developing countries for receiving and disbursing money. The actual funds contained in the LDCF are \$91.8 million; the SCCF \$59.9 million; and the SPA \$50 million (GEF 2008). This leaves almost \$100 million pledged to the UNFCCC that is still outstanding (Ayers 2009).

Second, many developing countries have expressed concern over the unclear guidance and high transaction costs attached to GEF funding mechanisms. For example, Möhner and Klein (2007) reviewed the fiscal and technical adequacy of the GEF managed funds and showed that the GEF does not provide adequate operational guidance making it difficult for developing countries to apply for project funding. Also, although funding through the GEF is not formally conditional, requirements attached to funding include burdensome reporting and co-financing criteria. For example, the LDCF and SCCF will only meet the costs of additional adaptation needs imposed on the country by the impacts of climate change, while the costs associated with baseline development activities (that would occur anyway in the absence of climate change) must be supported by co-financiers. However, as shown from the example of the Tuvalu NAPA, as mentioned above, distinguishing 'additional' costs of climate change impacts from baseline development needs is extremely complex if not impossible, while many countries cannot afford to meet the baseline development costs so the offer of funding for the additional cost is rather redundant.

Finally, funding through the GEF is disbursed via implementing agencies such as the UNDP, the UNEP and the World Bank, which adds further bureaucracy to the process. Many LDCs have expressed concern over the functional relationships with implementing agencies inhibiting access to funding for NAPAs from the LDCF. In the case of Tuvalu as mentioned above, the NAPA used UNDP Fiji as the Implementing Agency. In the course of the project, late disbursement of funds from UNDP led to delays in NAPA activities (Ayers and Huq 2008b).

Therefore, funding under the GEF is inadequate to meet the current costs of adaptation in developing countries, both fiscally and in terms of accessibility. The Adaptation Fund does provide a more positive model for funding under the UNFCCC. It has the potential to generate more significant sums, because it is funded by a 2% levy on Clean Development Mechanism (CDM) transactions, which could generate between $160–950 million (Müller 2007). There is also talk of applying the levy to international air travel, which has the potential to add $4–10 billion annually (Müller 2007). The Adaptation Fund also has a unique governance structure, and is not managed by the GEF but by an independent board with representation from the five UN regions as well as special seats for LDCs and SIDS. Countries can make submissions for funding directly to the Fund as opposed to going through designated implementing agencies. However, the Adaptation Fund is not yet operational, and even optimistic projections of the funds that could be raised do not meet estimates of the total costs of adaptation in developing countries.

Given the limitations in both size and scope of the UNFCCC funds for meeting the costs of adaptation, other sources of funding for adaptation that are increasingly being looked into include Official Development Assistance (ODA). In some ways, this makes a lot of sense, given that building adaptive capacity and reducing vulnerability must be underpinned by good development, and so requires actions that have traditionally been the focus of development practitioners.

As such, many donor agencies have begun to generate funds through bilateral and multilateral institutions for adaptation. For example, the World Bank has recently launched its Climate Investment Funds (CIFs). The overall objective of the CIFs is to provide concessional loans for policy reforms and investments that achieve development goals through a transition to a low-carbon development path, and a climate-resilient economy (World Bank 2008). One of the programmes under the CIFs relevant to adaptation is the Pilot Programme on Climate Resilience (PPCR), with a target size of $1 billion, aimed at increasing climate resilience and building adaptive capacity in developing countries.

However, the development of the CIFs in general, and the PPCR in particular, has not been without controversy (Müller 2008). The governance structure of the CIFs and PCCR have been heavily criticized as being donor-driven (CCCD 2008). Further, most of the funding under the PPCR will be made available through loans, not grants, and these loans are to be counted as ODA. This has been heavily criticized because financing for adaptation should be seen as compensation from high emission countries for those that are most vulnerable to climate change impacts, and so additional to existing aid commitments. This principle is specifically recognized by the UNFCCC through Article 4.4, which specifies that developed countries have committed to helping 'particularly vulnerable' countries meet the costs of adaptation. Funding adaptation under the UNFCCC meets this requirement, while funding adaptation through ODA, does not (Ayers 2009).

The PPCR was particularly open to such criticisms because it was originally entitled the Adaptation Pilot Fund, widely interpreted as a move to compete with the Adaptation Fund under the Kyoto Protocol (Müller 2008). The initiative was understandably ill received by the developing countries and the NGO community, particularly given that donor-instigated adaptation funds are counted as part of the 0.7% ODA commitments, which are not fulfilled by the majority of donors. Such an arrangement rings of earmarking aid for adaptation purposes and the imposition of 'new conditionalities', going against the principles of the Paris Declaration on Aid Effectiveness, including partner country ownership of aid budgets (Ayers and Huq 2008b). In response to these criticisms, the Bank redesigned the 'Adaptation Pilot Fund' to the 'Pilot Programme on Climate Resilience'. A more recent revised proposal included a more balanced representation of donor and recipient countries and includes the chair of the Adaptation Fund Board, in an attempt to remove the sense of competition with the Adaptation Fund and instead reframe the PPCR as a complementary to the UNFCCC process.

Although these arrangements are still under discussion, this does point to progress on understanding the role of ODA for contributing to broader adaptive capacity – or 'climate resilient development' – rather than specific and additional climate change adaptation. Rather than funding adaptation, as it is understood by the UNFCCC, and which is morally an additional responsibility of developed countries to existing

ODA commitments, new development funds relevant to climate change adaptation should be used to fund what the UNFCCC cannot; broader resilience building, necessary for 'additional' adaptation to be successful (Ayers and Huq 2008b).

Overcoming the challenge for raising funds for adaptation therefore lies in recognizing the separate but complementary roles of climate change and development institutions, and ensuring effective collaboration between these two fields to achieve the common goals of vulnerability reduction. There must be understanding not only of the implications of climate change for development, but, importantly, of the need for good development to underpin adaptation. Funding 'stand-alone' or discreet adaptation options cannot be effective without taking into consideration, calculating, and then meeting, the wider costs of development interventions that will ensure adaptation to be effective for the most vulnerable.

Doing adaptation

Within the international climate change arena, the preoccupation with estimating the costs for adaptation and raising the funds to meet those costs is underpinned by an assumption that if adequate funding available for the most vulnerable developing countries to adapt, then adequate adaptation could take place (Ayers 2009). However, this overlooks the fact that there is a significant deficit in 'absorptive capacity' (Müller 2008) of many developing countries and the most vulnerable groups within them to access these funds, and then to use them for their intended purposes. A major challenge for adaptation is ensuring that the money raised can be received and delivered in ways that facilitate the adaptation of the most vulnerable.

In terms of accessing funds, with the increasing proliferation of bi- and multi-donor and Convention funds available, it is vital to avoid duplication of efforts and ensure consistency in approach. Bangladesh has taken steps to address the problem of accessing the proliferation of international funds for adaptation, as well as to build national 'absorptive capacity' to receive these funds, through a Multi-Donor Trust Fund (MTDF) via which all funding for adaptation could be channelled (Box 2).

A further challenge is to ensure that, once accessed, careful consideration is given to the institutions and agencies through which

Box 2 Bangladesh Multi-Donor Trust Fund

The Multi-Donor Trust Fund (MDTF) currently being launched in Bangladesh will have money obtained from different national and multilateral climate change funds and be managed by a board of trustees. In Bangladesh's National Budget for 2008/09 the finance minister recently allocated approximately $45 million from the national exchequer towards the climate change strategy implementation and invited international donors to provide matching funds into the MDTF. To date, the UK's Department for International Development (DFID) has pledged £75 million in grant funding to initiate this fund; Denmark has pledged to contribute to the MTDF; and other countries are expected to follow suit. It is hoped that such a framework could significantly reduce transaction costs for global and bilateral funds and pave the way for large fund flows in the future, while ensuring proper institutional structures, governance, management, and targeting of funds at the national level (Ayers, Alam, and Huq 2008)

The scale and scope of the MDTF are still under discussion at the time of writing, but it is hoped that the fund would be accessible to the government agencies, NGOs and private sector actors who work with vulnerable communities at the local level. However, a recent announcement that the World Bank is likely to take at least interim (two years) responsibility for the governance of the fund has raised alarm bells, particularly from NGOs concerned that this may result in bureaucratic and functional problems similar to those already faced with the international, and in particular the GEF-managed, adaptation funds. The Government of Bangladesh has also been criticized by civil society stakeholders for insufficient local-level participation in the development of the MDTF, and for allowing processes to be donor led (Ayers, Alam, and Huq 2008). It is important that the government responds to these criticisms and ensures adequate participation of civil society and local stakeholders as they take forward both the Climate Change Strategy and Action Plan and MDTF. Discussions concerning MDTF governance remain ongoing (Ayers 2009).

funds are channelled to the people most at risk from climate change. Given that those most vulnerable to climate change are the poorest people in developing countries, for adaptation funding to be effective in reducing vulnerability these groups must be targeted by adaptation resources. Adaptation to climate change at this level requires local knowledge, local competence. and local institutional capacity

(Sattherthwaite, Huq, Pelling, *et al.* 2007). However, providing governments with funding for adaptation does not necessarily mean that it will reach the most vulnerable, particularly given that these communities tend to be among the poorest and most marginalized groups in regions that are difficult to access and already receive limited government support (Huq and Reid 2007).

One approach that specifically targets vulnerable communities is Community Based Adaptation (CBA), encompassing collective NGO action on adaptation at the community level. CBA aims to identify, assist, and implement community-based development activities, research and policy, in regions where adaptive capacity is as dependent on livelihood indicators as climatic changes. It focuses on the ways in which interactions between governance, economic well being, health, education, equality, security, a healthy environment, and climate change can contribute to the vulnerability of individuals, households, and communities (Thompkins, Nickolson-Cole, Hurlston, *et al.* 2005). In practice, therefore, CBA is often not dissimilar to development (Polak 2007); the difference lying less with the intervention itself than the inputs to and motivations for the intervention. The adaptation element introduces the community to the notion of climate risk and then factors this risk into activities (Huq and Reid 2007) (Box 3).

The CBA movement is growing in size and significance and now includes advocates from major international NGOs (for example Oxfam, Practical Action, and the Red Cross/Red Crescent Societies), research institutions, academics, and local NGOs in vulnerable countries, and well as support from some large donor institutions. However, there are institutional barriers to the 'scaling up' of CBA initiatives, at least partially because of a lack of coherence across policy scales about what constitutes adaptation (and therefore what can justifiably be fiscally and institutionally supported by climate change frameworks) and how this is related to development (Ayers and Dodman 2009). Given that the success of adaptation at the local level relies heavily on supportive institutional environments created by the levels above, successful adaptation requires both the involvement of local groups and civil society organizations with the knowledge and capacity to act, and a willingness among local and national governments to work with and support adaptation among low-income and vulnerable groups.

Achieving wider institutional support for adaptation can be

Box 3 Community-based adaptation in Cavite City, the Philippines

Cavite City in the Philippines is situated on a peninsula surrounded by three bodies of water, with about half its population on the coast. Cavite experiences an average of two tropical cyclones every year, and the city is also affected by drought and sea-level rise, all of which are predicted to be greatly exacerbated by the impacts of climate change. Currently 10% of the population is vulnerable to sea-level rise, but a one-metre increase would put around two-thirds of the population at risk. Other threats linked to climate change include erosion, sedimentation, flooding, and saltwater intrusion into groundwater. Poor people are most at risk due to their vulnerability to climatic events and social, economic, technological, and institutional factors.

The community has responded through various autonomous and supported adaptation strategies, including: accommodating sea-level rise by building houses on stilts; strengthening/reinforcing the physical structure of houses; moving to safer places during extreme weather events; placing sandbags along the shorelines; and engaging in alternative income generating activities locally or in other areas, or changing occupation. However, these are not adequate to deal with the scale of the problem and are not sufficiently supported by wider development plans. Some government strategies (such as relief assistance, resettlement, and shoreline protection) have reduced the vulnerability of coastal households, but the measures are inadequate and costly.

Consultations revealed that communities feel they only have poor to fair human, physical, and financial capacities to face the threat of climate change. People expressed significant concern over climate risks, and proposed several adaptation strategies, many of which were community-based, non-structural, capacity-building measures. They include the following.

- Improved knowledge management such as community-based monitoring of changes in coastal areas for input into vulnerability and adaptation assessments; creation of community early warning systems; and documentation, sharing and promotion of traditional knowledge, skills and practices that enhance adaptation.
- Policy and institutional reforms: such as developing an integrated coastal zone management plan that includes land and sea use zoning, alternative livelihood development and eco-waste management; and providing secure property rights and micro-finance/insurance schemes that enhance the adaptive capacity of vulnerable groups.
- Capacity development: through raising awareness; participatory risk and adaptation assessment and planning; alternative livelihood development and creating a multi-sectoral integrated coastal zone management body.

Source Huq and Reid 2007

problematic in the most vulnerable developing countries where development needs must be the overriding priority of any national government. While development is likely to equate with vulnerability reduction, this is not always the case; such as, where economic development strategies increase dependency on climate-sensitive resources; or where adaptation interventions adopted by donors do not equate with the development priorities of recipient countries.

Achieving synergies among adaptation and development priorities is a major challenge for adaptation, particularly if ODA is being used to assist climate resilient development. One way of ensuring consistency in approach is to 'mainstream' climate change into other national development plans and priorities. Mainstreaming involves integrating climate change adaptation into social, institutional, and infrastructural development planning, and has been adopted in a variety of forms in development policy and practice. Mainstreaming is seen as making more sustainable, effective, and efficient use of resources than designing and managing policies separately form ongoing activities (Klein, Schipper, and Dessai 2003).

Klein (2008) discusses two types of mainstreaming in development; first, a 'technology-based view of adaptation' that sees mainstreaming in terms of ensuring that projections of climate change are considered in the decision-making of relevant government departments and agencies, so that technologies are chosen that are suitable for a future climate, 'climate proofing' development. Second, a 'development-based view of adaptation', which ensures that in addition to climate proofing, development efforts are consciously aimed at reducing vulnerability by including priorities that are essential for successful adaptation, such as ensuring water rights to groups exposed to water scarcity during a drought. This latter option takes a more holistic approach to adaptation, seeing responses not as stand-alone or discreet climate-specific options, but as also addressing the underlying drivers of vulnerability that expose people to climate change impacts. The 'development-based' approach to mainstreaming ODA provides an opportunity to facilitate mainstreaming across national development planning and processes in partner countries (Klein 2008).

However, while there is a lot of rhetoric about integrating adaptation into development, there is very little understanding of exactly what this means in practice (Ayers and Huq 2008b). To echo calls from a recent WRI report in relation to mainstreaming, 'concrete models and approaches are needed' (Bapna and McGray 2008) presenting a further challenge for achieving adaptation based on sustainable development. Any such model must begin with raising awareness across all scales of the implications of climate change for existing development concerns, and draw on local experiences of how the risks of climate change can impact development. It is important that tools for climate change data analyses are adapted to provide information that is meaningful and relevant to the everyday risks that people experience. This requires downscaling climate modelling data as far as can be useful (currently to the regional level), while supporting and maintaining climate observation stations at local and regional scales to relate climate information to observed climatic changes. Improving climate services, raising awareness about available climate information, and providing evidence of its value to decision makers, are all essential to aligning development and climate change priorities at higher scales.

Conclusions

Although good progress has been made on adaptation over the last decade, with the last few years in particular seeing adaptation rise up the climate change agenda going from marginal to prominent policy option, many challenges remain. First, there persists a great deal of confusion over what adaptation actually is, particularly in terms of where adaptation interventions lie in relation to development. 'Stand-alone' adaptation to climate change will not be effective in building the adaptive capacity of the most vulnerable, given that climate change vulnerability is determined by factors related to development. Greater attention is needed to recognizing the links between development and adaptation, and exploring the synergies between the two fields in practice. Adaptation must be based on good development, and development must be climate resilient in order to be sustainable in a changing climate.

Second, the costs of adaptation are already and will continue to be significant and well in excess of development budgets and ODA for the most vulnerable countries. While the establishment of the existing funds for adaptation, and the recent progress on the adaptation fund in particular, are commendable, these funds do not amount to anywhere near enough, and high-income nations must accept responsibility for the causes of climate change and step up efforts for funding adaptation in developing countries. In addition, more must be done to make existing funds more accessible. At the same time, the role of ODA in financing adaptation needs to be clarified, to ensure that the principle that funding for adaptation is additional to aid is upheld, while acknowledging that there is much that donors can do to assist in building adaptive capacity in developing countries.

Finally, action on implementing adaptation must be coupled with efforts to fully integrate adaptation into development planning, to ensure that the trade-offs between adaptation and development are minimized, and maladaptive actions that enhance rather than reduce vulnerability to the impacts of climate change, are avoided. Climate change requires a shift from business-as-usual development, to climate resilient development, in such a way that guarantees that the underlying objectives of sustainable development are not compromised. As noted by Schipper (2007), focusing on adaptation before aligning development processes with wider aspects of vulnerability reduction, is 'putting the cart before the horse'.

Given that climate change is already happening, and the impacts are beginning to be felt, it is important that action on adaptation is not delayed. A 'learning by doing' approach to adaptation, particularly at the community level, is essential to start building the resilience of the most vulnerable. Yet, urgency on adaptation must not overshadow the fact that many challenges in adaptation remain, which must be given careful consideration in order to build resilience to climate change, sustainably.

References

Action Aid. 2007
Compensating for Climate Change: principles and lessons for equitable adaptation funding
Washington, DC: Action Aid

Agrawala S and Fankhauser S (eds). 2008.
Economic aspects of adaptation to climate change: costs, benefits and policy instruments
Paris: Organization for Economic Co-operation and Development (OECD), 133 pp.

Agrawala S, Mohner A, Hemp M, van Aalst M, Hitz J, Smith H, Meena S, Mwakifwamba S M, Hyera T, and Mwaipopo O U. 2003
Development and Climate Change in Tanzania: focus on Mount Kilimanjaro
Paris: Organization for Economic Co-operation and Development (OECD)

Ayers J. 2009
Financing urban adaptation
In *Adapting Cities to Climate Change*, edited by J Bicknell, D Dodman, and D Satterthwaite
London: Earthscan

Ayers J and Dodman D. 2009
Climate Change Adaptation and Development: the state of the debate
Accepted for publication by Progress in Development Studies

Ayers J and Huq S. 2008a
The value of linking mitigation and adaptation: a case study of Bangladesh
Environmental Management
[Published online: 28 October 2008 doi:10.1007/s00267-008-9223-2]

Ayers J and Huq S. 2008b
Supporting adaptation through development: what role for ODA?
Paper presented at Development's Invisible Hands: development futures in a changing climate, DSA Annual Conference, London, November 2008.

Bapna M and McGray H. 2008
Financing Adaptation: opportunities for innovation and experimentation
Washington, DC: World Resources Institute

Burton I, Diringer E, and Smith J. 2006
Adaptation to Climate Change: international policy options
Arlington, VA: PEW Centre on Global Climate Change.

Dodman D, Ayers J M, and Huq S. 2009
Building resilience. State of the World 2009: into a warming world
Washington, DC: Worldwatch Institute

GEF (Global Environment Facility). 2008
Status Report on the Climate Change Funds as of March 4th, 2008 (Report from the Trustee)
Washington, DC: GEF

Huq S and Ayers J. 2007
Critical list: the 100 countries most vulnerable to climate change
London: International Institute of Environment and Development (IIED)
[IIED Opinion Paper]

Huq S and Ayers J. 2008a
Climate change impacts and responses in Bangladesh
Briefing note prepared for the European Parliament, Policy Department Economic and Scientific Policy, DG Internal Policies of the Union (Brussels).

Huq S and Ayers J. 2008b
Taking steps: mainstreaming national adaptation
London: International Institute of Environment and Development (IIED)
[IIED Briefing Paper, November 2008]

Huq S and Reid. 2007
Community based adaptation
London: International Institute of Environment and Development (IIED)
[IIED Briefing Paper]

Huq S, Reid H, and Murray L A. 2006
Climate change and development links
London: IIED (Gatekeeper Series 123)

IPCC (Intergovernmental Panel on Climate Change). 2001
Climate Change 2001: impacts, adaptation and vulnerability
Cambridge: Cambridge University Press
[Contribution of Working Group II to the 3rd Assessment Report of the IPCC].

IPCC (Intergovernmental Panel on Climate Change). 2007
Summary for policymakers
In *Climate Change 2007: impacts, adaptation and vulnerability*, edited by M L Parry, O F Canziani, J P Palutikof, P J van der Linden, and C E Hanson
Cambridge: Cambridge University Press, pp. 7–22
[Contribution of Working Group II to the Fourth Assessment Report of the Intergovernmental Panel on Climate Change]

Klein. 2008
Mainstreaming climate adaptation into development policies and programmes: a European perspective
In *Financing Climate Change Policies in Developing Countries*
[European Parliament, Brussels]. PE 408.546 -IP/A/CLIP/A/CLIM/ST/2008-13

Klein R J T, Schipper E L, and Dessai S. 2003
Integrating mitigation and adaptation into climate and development policy: three research questions
East Anglia: Tyndall Centre for Climate Change Research
[Tyndall Centre Working Paper 40]

Klein R J T, Huq S, Denton F, Downing T E, Richels R G, Robinson J B, Toth F L. 2007
Inter-relationships between adaptation and mitigation
In *Climate Change 2007: impacts, adaptation and vulnerability*, edited by M L Parry, O F Canziani, J P Palutikof, P J van der Linden, and C E Hanson
Cambridge: Cambridge University Press, pp. 745–77
[Contribution of Working Group II to the Fourth Assessment Report of the Intergovernmental Panel on Climate Change]

Klein R and Persson A. 2008
Financing Adaptation to Climate Change: issues and priorities
[ECP Report No. 8. October 2008, Brussels]

Levina E (2007)
Adaptation to Climate Change: international agreements for local needs
Document prepared by the OECD and IEA for the Annex I Expert Group on the UNFCCC. OECD/IEA 2007, Paris
[Details available at <www.oecd.org/env/cc/aixg>, last accessed in December 2008]

McGray H, Hammill A, and Bradley R. 2007
Weathering the Storm: options for framing adaptation and development
Washington, DC: World Resources Institute

Möhner A and Klein R J T. 2007
The global environment facility: funding for adaptation or adapting to funds?
Stockholm: Stockholm Environment Institute. 26 pp.
[Climate and Energy Programme Working Paper]

Müller B. 2007
The Nairobi climate change conference: a breakthrough for adaptation funding
Oxford Energy and Environment Comment
[Oxford Institute for Energy Studies]

Müller B. 2008
International adaptation finance: the need for an innovative strategic approach
Oxford: Oxford Institute for Energy Studies
[Background Policy Paper for the Climate Strategies Project on Post-2012 Policy Framework: Options for the Tokyo G8 Summit]

Oxfam International. 2007
Adapting to climate change: what's needed in poor countries, and who should pay
Washington, DC, Brussels, Geneva, and New York: Oxfam
[Oxfam Briefing Paper **104**]

Polak E. 2007
South Asian Voices on Climate Change: Investigating the value and contribution of local knowledge for community-based adaptation to climate change
Practical Action
Details available at <www.practicalaction.org>

Satterthwaite D, Huq S, Pelling M, Reid H, Lankao R P. 2007
Adapting to climate change in urban areas: the possibilities and constraints in low- and middle-income nations
London: International Institute of Environment and Development (IIED)
[Human Settlements Discussion Paper Series – Climate Change and Cities **1**]

Scheuren J M, le Polain de Waroux O, Below R, Guha-Sapir D, Ponserre S. 2008.
Annual Disaster Statistical Review: the numbers and trends 2007
Belgium: Université Catholique de Louvain, Center for Research on the Epidemiology of Disasters (CRED)

Schneider S H, Semenov S, Patwardhan A, Burton I, Magadza C H D, Oppenheimer M, Pittock A B, Rahman A, Smith J B, Suarez A., Yamin F. 2007
Assessing key vulnerabilities and the risk from climate change
In *Climate Change 2007: impacts, adaptation and vulnerability*, edited by M L Parry, O F Canziani, J P Palutikof, P J van der Linden, and C E Hanson
Cambridge: Cambridge University Press, pp. 779–810
[Contribution of Working Group II to the Fourth Assessment Report of the Intergovernmental Panel on Climate Change]

Schipper L. 2006
Conceptual history of adaptation in the UNFCCC process
RECIEL **15**: 1

Schipper L. 2007
Climate change adaptation and development: exploring the linkages
Norwich, UK: Tyndall Centre for Climate Change Research, University of East Anglia
[Tyndall Centre Working Paper Series **107**]

Smit B and Wandel J. 2006
Adaptation, adaptive capacity and vulnerability
Global Environmental Change **16**: 282–292

Smit B, Burton I, Klein R J T, Wandel J. 2000
An anatomy of adaptation to climate change and variability
Climatic Change 45(1): 223–251

Thompkins E L, Nickolson-Cole S A, Hurlston L-A, Boyd E, Brooks Hodge G, Clarke J, Gray G, Trotz N, Varlack L. 2005
Surviving Climate Change in Small Islands – a guidebook
Norwich, UK: Tyndall Centre for Climate Change Research, University of East Anglia

UNDP (United Nations Development Programme). 2007
Human Development Report 2007/2008: fighting climate change: human solidarity in a divided world
New York: UNDP

Wisner B, Blakie P M, Cannon T, Davis I. 2003
At Risk: natural hazards, people's vulnerability and disasters, 2nd edn
London: Routledge

World Bank. 2006
Clean Energy and Development: towards an investment framework ESSD-VP/I-VP
Washington, DC: World Bank

World Bank. 2008
Q&A: climate investment funds
Washington, DC: World Bank

4

Research and development priorities for climate change mitigation and adaptation

Thomas J Wilbanks*

By far the largest share of support for research on climate change over the past three decades has been focused on atmospheric and earth system science, because the first questions for climate change discussions were whether climate change is really happening, what earth system changes would be associated with it, and how it is related to net fluxes of greenhouse gases (GHGs). Research support for understanding responses to climate change has historically been far more limited, for at least two reasons in the early decades: (a) it was believed that, if climate change is real, its dangers are clearly so alarming that GHG emissions must be stabilized quickly; research may be needed to understand impacts and mitigation options, but we will know enough about appropriate responses that are needed for research on impacts of climate change are not likely to be extensive; and (b) it was believed that looking into adaptation to impacts as a response alternative would reduce the necessary insistence on mitigation in order to avoid impacts.

Over the past decade, however, perspectives on the need for new knowledge have changed profoundly, although patterns of research and development (R&D) support have been slow to adjust. Because impacts of climate change are already being observed (IPCC 2007a), they must be analysed and adaptations must be considered. Because, physical impacts seem to be emerging more rapidly than expected even a decade ago and because carbon emissions are rising more rapidly

---

* Oak Ridge National Laboratory, USA.

than projected by any climate change scenario that has been considered by the IPCC (Raupach 2007), the central question for climate change science and policy has shifted from ‘whether’ to ‘so what?’ How much difference is it likely to make? How serious are the risks and vulnerabilities? What does this tell us about how we should respond?

Over the past years, because of the scarcity of support for research on the causes of climate change, because of the even more severe scarcity of support for research on impacts, and because of the near-absence of support for research on adaptation, the climate science community has had a very limited ability to answer many of these questions. This is no excuse for failing to act to reduce GHG emissions and to enhance resilience with respect to climate change risks and vulnerabilities, but it focuses attention on the importance of targeted support for R&D related to climate change mitigation and adaptation (M&A).

Quite a number of studies have considered aspects of this need, and because the needs are so extensive – compared with a strong research base in other aspects of climate science – the research gaps are extraordinarily wide-ranging. This paper will survey the landscape of discussions of research needs as it exists at the moment, based on many sources. Because the landscape is so rich and diverse, each section will end with a summary of what appear to be the most important places to start.

This paper will consider the needs and priorities for R&D related to climate change M&A and relationships between them, including technology development, and it will also consider issues in moving that new knowledge into action. It is important to recognize, however, that needs for R&D are not excuses for failing to take climate change actions now, for at least two reasons. The most important is that the risks and vulnerabilities associated with climate change, combined with the long-term significance for the global climate of what we do in the near term, tell us that we cannot afford to wait to act. While we learn more, there are a host of actions that we already know make sense. The second reason is that acting and learning benefit from linkages with each other. They can and should move ahead in parallel and in partnership, each learning from the other, on an adaptive response pathway.

Mitigation and adaptation research and development

Research and development needs and priorities include mitigation R&D, adaptation R&D, and challenges in developing response strategies that integrate M&A as parts of a coherent integrated response to concerns about climate change.

Climate change mitigation R&D

In some ways, mitigation policy-making has proceeded ahead of the development of knowledge bases to assure that its initiatives will work, which may be one of the many reasons why progress has often been so slow. The continuing tension is between the need for better knowledge but the even more pressing is the need to take actions before long periods of time pass, so that knowledge improvement and policy-making tend to occur in parallel rather than in series, even if that is not what most scientists would consider the ideal.

The first and most cross-cutting mitigation R&D need, then, is to treat ongoing mitigation policy as a host of 'natural experiments' that are rich opportunities for learning about how changes can be induced and facilitated. Mitigation actions should be paired with systems for observing and analysing the experience, so that subsequent policies are likely to work better.

More generally, climate change mitigation R&D priorities include efforts to improve the understanding of policy options; to improve data collection, modelling, and analysis; to improve the basic foundations for knowledge development and application; and to improve mitigation technologies and/or reduce their costs. These R&D priorities are described below.

Research to improve the understanding of policy options Four especially high-priority research needs have been identified, namely non-climate policies affecting GHG sources and sinks, alternative development paths and their implications for reducing/avoiding GHG emissions, converting mitigation capacities into mitigation actions, and processes of decision-making (IPCC 2007b, ch. 12.4). To this list might be added realistic avenues to enable major developing economies to reduce their rate of growth in GHG emissions without undermining their sustainable development (Wilbanks 2008a).

Research to improve data collection, modelling, and analysis related to climate change mitigation Again, a number of important generic needs have been identified, including reliable emission data sets and projections (for example, treatment of non-CO_2 GHGs); links between climate change and other policies (for example, sustainable development); and studies of costs and potentials, reducing serious constraints on useful analysis at present, such as combining top-down and bottom-up elements, knowledge about spill-over effects (on other sectors or countries), technological change, tradeoffs between mitigation actions and other activities, such as in the case of biomass (see below regarding M&A synergies and tradeoffs), and relationships between rates of adoption of mitigation technologies and other influences on human and institutional behaviour (IPCC 2007c, ch. 14).

Research to strengthen the scientific basis of integrated assessment models that are vitally important tools for examining implications of different pathways for technology and policy choices related to GHG emissions For example, the central role of these integrated assessment models in providing emission assumptions for the new family of climate change scenarios currently being developed for future use by the IPCC (IPCC Expert Meeting Report 2008). Recent discussions of science needs (for example, DoE 2008) have identified research needs related to such issues as connections with impact and adaptation knowledge, estimating and representing uncertainties, pros and cons of inclusiveness versus agility, incorporating governance/ institutional behaviour as a variable affecting outcomes of processes being modelled, and experimenting with multidimensional/multi-metric valuation of costs and benefits.

Research to improve basic foundations for knowledge development and application Attention is also being paid to the fact that, in many cases, relatively applied mission-oriented mitigation (and adaptation) research is undermined by weaknesses in fundamental disciplinary research. A recent report by the US National Research Council for the US Climate Change Science Program identified eight high-priority research needs, five substantive and three cross-cutting (CHDGC 2007). According to this report, substantive research priorities are improving the understanding of environmentally significant consumption, improving the fundamental understanding of risk-related judgement and decision-making under uncertainty, improving the understanding of how social institutions affect resource use, improving the understanding of socio-economic change as context for climate change impacts and responses (especially technological change and land-use change), and improving understandings of the valuation of climate consequences and policy responses. Cross-cutting priorities are observations, indicators, and metrics; non-linearities, feedbacks, and thresholds in system responses to climate change in a multi-causal setting; and scale dependencies and cross-scale interactions.

R&D to improve mitigation technologies and/or reduce their costs
Finally, R&D is needed to improve the portfolio of technology options available for achieving mitigation targets. The fundamental challenge, of course, is to reduce net emissions of GHGs (considering both sources and sinks) while at the same time providing energy and other services to support economic and social development. We know two things for certain at this point: that the current portfolio of technologies is not adequate to meet this challenge and that the net benefit/cost calculus of mitigation depends on success with technology development and deployment (Interlaboratory Working Group 2000).

Elements of the challenge include reducing emissions from energy production (for example, in energy conversion facilities), reducing emissions from energy consumption (for example, in industry, transportation, and buildings), and enhancing uptakes of carbon by sinks (growing biomass, soils, and the ocean). Options tend to be distinguished by their element and their time scale (for example, near, mid, and long term).

In very general terms, the challenges calling for technology R&D contributions include the following.

- *Renewable energy alternatives* This involves reducing costs relative to fossil energy alternatives and realizing potentials for large-scale contributions, such as providing most of the energy for a major urban-industrial complex.
- *Nuclear energy alternatives* . This involves addressing challenges of the nuclear fuel cycle, such as radioactive waste transport and disposal, and its connections with global security issues.
- *Decoupling fossil energy sources from carbon emissions* This also involves focusing on technologies for carbon capture and sequestration (storage).

- *Increasing the efficiency with which energy is used and moving towards energy sources that involve significantly less carbon emissions* This is a special challenge for the transportation sector, which is especially dependent on liquid fossil fuels.
- *Moving towards truly new technology alternatives, rather than simply assuring incremental improvements in known technologies* This is necessary in order to transcend inherent limitations in the capabilities of known technologies to meet global needs for energy services.

Table 1 summarizes the conclusions of one relatively comprehensive study of technology R&D priorities (CCTP 2006). According to this study, focused on the relatively near term, the highest priorities include transportation fuel cell systems, improved renewable energy technologies, hydrogen storage, higher-efficiency fossil energy technologies, and affordable technologies for carbon sequestration. The CCTP report also identifies very useful technology requirements for monitoring GHG emissions and verifying claims of GHG emission reduction.

IPCC's *Fourth Assessment Report* (IPCC 2007b, Ch. 11) identifies five 'cross-sectoral' technologies that merit special R&D emphasis, including information technologies (which have the potential to revolutionize information collection, storage, analysis, and communication), carbon capture and storage, hydrogen, biomass energy, and geo-engineering.

Among all of the items, the most important priorities for mitigation R&D are as follows.

- Establishing the feasibility and affordability of carbon capture and sequestration on a large scale (if possible).
- Analysing the ongoing climate change mitigation policies as natural experiments.
- Understanding relationships between mitigation and sustainable development.
- Improving the understanding of consumption behaviour.

Climate change adaptation R&D

Even though research is needed to improve the understandings of climate change mitigation, knowledge about climate change adaptation – the avenues, their potentials, their costs and benefits, their possible limits – is in fact far more limited, to the obvious distress of decision-

Table 1 Climate change technology development and deployment for the 21st century

Goal	Near-term	Mid-term	Long-term
Energy end-use and infrastructure	• Hybrid and plug-in hybrid electric vehicles • Engineered urban designs • High-performance integrated homes • High-efficiency appliances • High-efficiency boilers and combustion systems • High-temperature superconductivity demonstrations	• Fuel cell vehicles and H_2 fuels • Low-emission aircraft • Solid-state lighting • Ultra-efficient HVACR • Smart buildings • Transformational technologies for energy-intensive industries • Energy storage for load leveling	• Widespread use of engineered urban designs and region planning • Energy managed communities • Integration of industrial heat, power, process, and techniques • Superconducting transmission and equipment
Energy supply	• IGCC commercialization • Stationary H_2 fuel cells • Cost-competitive solar PV • Demonstrations of cellulosic ethanol • Distributed electric generation • Advance fission reactor and fuel cycle technology	• FutureGen scale-up • H_2 Co-production from coal/biomass • Low wind speed turbines • Advanced biorefineries • Community-scale solar • Gen IV nuclear plants • Fusion pilot plant demonstration	• Zero-emission fossil energy • H_2 and electric economy • Widespread renewable energy • Bio-inspired energy and fuels • Fusion power plants
Capture storage and sequestration	• CSLF and CSRP • Post combustion capture • Oxy-fuel combustion • Enhanced hydrocarbon recovery • Geologic reservoir characterization • Soils conservation • Dilution of direct injected CO_2	• Geological storage proven safe • CO_2 transport infrastructure • Soils uptake and land use • Ocean CO_2 biological impacts addressed	• Track record of successful CO_2 storage experience • Large-scale sequestration • Carbon and CO_2-based products and materials • Safe long-term ocean storage
Other gases	• Methane to markets • Precision agriculture • Advances refrigeration technologies • PM control technologies for vehicles	• Advanced landfill gas utilization • Soil microbial processes • Substitutes for SF_6 • Catalysts that reduce N_2 to elemental nitrogen in diesel engines	• Integrated waste management system with automated sorting, processing and recycle • Zero-emission agriculture • Solid-state refrigeration/AC system
Measure and monitor	• Low-cost sensors and communications	• Large-scale, secure data storage system • Direct measurement to replace proxies and estimators	• Fully operational integrated MM systems architecture (sensors, indicators, data visualization and storage, models

Note HVACR – heating, ventilating, air conditioning, refrigeration; IGCC – integrated gasification and combined cycle technology; CSLF – Carbon Sequestration and Leadership Forum; CSRP – Carbon Sequestration Regional Partnerships
Source CCTP (2006)

makers suddenly asking for advice regarding adaptation. Most of the adaptation research reported by the IPCC's Third and Fourth Assessment Reports is focused on adaptation capacity, that is, determinants about the ability to adapt if a party decides it wants to. It says relatively little about what actions should be considered in order to manage climate change risks through adaptive behaviour, except for a few topics such as economic costs and benefits of building dikes to protect seacoasts threatened by sea-level rise (Tol 2002).

The biggest challenge for adaptation knowledge development is that 'one size does not fit all'. In many cases, the most promising adaptation actions are likely to differ between scales, locations, sectors, and even potentially affected populations. Moreover, portfolios of possible adaptation actions will mix institutional and technological alternatives (Wilbanks, *et al.* 2007).

For the time being, there is little evidence of successful adaptations to climate change, because changes that can be attributed to climate change are still limited, and most of the modest number of adaptation actions have been taken so recently that their effects are difficult to evaluate. As a result, most of the current knowledge about adaptation is drawn from experience, adapting to climatic variations, such as annual and seasonal fluctuations in water scarcity, or from other analogues in the history of adaptation to major environmental changes. On the other hand, if climate change takes climate-related variables outside the range of historical experience, options that have been effective in addressing variation may not be sufficient for addressing change; in fact, confidence based on success in coping with variability can in some cases reduce sensitivity to risks from climate change.

Given the very limited current base of knowledge, research needs are enormous. One workshop held at the US National Academies on 25 October 2007 concluded that the following questions are especially important.

- What do we know about climate change and its impacts? (for example, observations, monitoring, impact metrics)
- What are the costs, benefits, and limits to adapting to climate change? (for example, uncertainties; adaptive capacity and resilience of natural, built, and human systems; reactive adaptation versus anticipatory adaptation; ethical, economic, and institutional issues)
- What are the most pressing and/or common needs for adaptation planning, and what strategies exist to prioritize among these

needs? (that is, information, tools, techniques, resources, and laws; relationship between adaptation and mitigation; common versus unique need of different sectors and at different spatial scales)

- What do we know about adaptation strategies and options, and what do we need to know to adapt more effectively? (that is, current best practices and illustrative future scenarios of effective adaptation efforts, including observations/monitoring, assessments and other decision support tools, communication strategies, and metrics/measures of success)
- What are the appropriate roles for governments and other entities in making investments, meeting needs, and setting adaptation priorities, and how can all of these entities be made to work together to effectively address issues that span multiple sectors and spatial and temporal scales? (climate services, decision support, responsibilities for all aspects of adaptation)
- What else do we need to know about adaptation to climate change? (for example, resilience to multiple stresses; potential to adapt to unexpected/abrupt changes; research priorities in natural sciences, social sciences, applied sciences, and technology development)

In summary, research and development needs for climate change adaptation include several aspects of responses—analysis and assessment, infrastructure development, and technology development (Wilbanks 2007).

- Analysis and assessment involves improving the ability to evaluate adaptation alternatives—evaluations of potentials, benefits and costs, and possible limits of adaptation strategies and options; risk assessment and management; financial incentives for increasing resiliency; vulnerability and hazard mapping; multi-stressor analysis (climate change impacts in a larger context); and improved climate change projections at detailed scales.
- Infrastructure development is focused on institutional responses that add resilience to possible climate change, such as emergency preparedness and response (including early warning systems), infrastructures for moving water from surplus to deficit areas, (including inter-basin planning and coordination), assistance programmes for especially vulnerable groups, land-use planning in vulnerable areas, and stronger health care systems in areas of vulnerability.

- Technology development for climate change adaptation is highly dependent on the affected sector and the time frame, but illustrations include biomass species development (for example, heat or drought-resistant varieties), affordable desalination, more efficient water-consuming technologies, from power plant cooling to water end uses, and technologies for critical area protection without unintended consequences.

The following two sectoral cases may be considered as examples.

(1) *Cities and other human settlements (CCSP 4.6, 2008):* Cities vary greatly by region, size, age, and other characteristics, but vulnerabilities include health, water and other urban infrastructure, effects of extreme severe weather events, energy requirements, the urban metabolism (for example, heat island effects), economic competitiveness, social and political structure (additional stresses and costs), and sea-level rise. Some of these can have positive effects in some areas as long as climate change is not too severe.

The major categories of adaptation options, all of which need research to assess potentials, benefits and costs, and possible limits, include

- changing the location of people or activities;
- changing the spatial form of a settlement (for example, green buildings, green spaces);
- technological change to reduce the sensitivity of infrastructures (for example, more affordable space cooling, surface materials that reduce heat island effects, waste reduction and treatment, warning systems);
- institutional change to improve coping capacities; and
- 'no regrets' approaches that add resilience at very low net costs.

More fundamental research needs include

- more research on climate change effects on US settlements, including in-depth studies of impacts in vulnerable regions;
- better projections of climate change at urban-area scales; improving the understanding of relationships between settlement patterns and resilience/adaptation;
- improving technologies that can reduce impacts as well as the vulnerabilities of buildings, urban infrastructure (for example, coping with warming, water scarcity, storm exposure, sea-level rise); and

- improving the understanding of how to encourage decentralized adaptive behaviour.

(2) *Energy production and use (CCSP 4.5, 2007):* Adaptation research needs for this sector are both relatively general and relatively technology-specific, involving improvement in knowledge of such issues as the following.

The general needs include

- implications of extreme weather events for energy system resilience, including strategies for reducing and recovering from impacts;
- technologies and practices that save cooling energy and reduce electrical peak loads, especially coping with heat waves;
- improvements in understanding the effects of changing energy resource/technology portfolios on regional energy balances and economies;
- likely effects in Arctic regions and on the intensity of storms to guide technology applications and development, including energy exploration and development; and
- linkages between climate change adaptation and targets for GHG stabilization levels.

The technology-specific needs refer to

- more efficient and affordable space-cooling technologies;
- improved information about regional energy implications of water scarcity, along with alternative adaptive responses;
- strategies and technology applications to increase the resilience of coastal and off-shore oil and gas production and distribution systems to severe weather events;
- strategies and technology applications to improve resilience of electric supply systems that may be subject to stress, for example, through greater tolerance for heating, expanded regional inter-tie capacities, and more distributed generation; and
- improved understanding of implications for energy production from wind, solar, and biomass technologies.

Among a great many pressing research needs, however, the highest priorities appear to be the following.

- Improving the knowledge of potentials, costs and benefits, and possible limits of strategies for adapting to impacts of climate change that are of particular concern.
- Improving the understanding of how to encourage and facilitate adaptation, especially interactions between top–down sources of financial and other support and bottom–up actors who need empowerment and flexibility to determine what makes sense for their contexts and act on those understandings, without a discouragingly heavy burden of bureaucratic requirements from larger scales.

R&D to support integrating mitigation and adaptation

The new priority is to improve the capacity to think about M&A not as separate kinds of responses to climate change but as parts of an integrated response portfolio (Wilbanks, *et al.* 2003). Integration of M&A is vastly complicated by the fact that their communities of knowledge and action, as well as their costs and benefits, are so different. For example, mitigation pathways show higher net benefits at a global scale than a local scale, because so many benefits from mitigation investments are external to a local area, while adaptation pathways show higher net benefits at a local scale, at least in industrialized countries, because significant potential impacts at a global scale are difficult to address through adaptation.

Comparisons, however, also depend on the magnitude and rate of climate change assumed in modelling scenarios. In fact, M&A are not necessarily competitors in climate change impact response strategies. In many cases, they are complements. If mitigation is successful in keeping climate change impacts to a moderate level, then adaptation can handle a larger share of the resulting impacts. Also, a GHG stabilization level that avoids dangerous impacts for many affected parties can be higher in a highly adaptable world than a poorly adaptable world.

What is then the right balance if leadership and investment resources are likely to be limited? Many M&A actions make obvious sense now, often because they offer considerable co-benefits. But in many other cases, effective (and persuasive) integration needs better information, better capacities for analysis and action, and further policy-making (Wilbanks and Sathaye 2007).

- *Needs for information* For broad-based quantitative analysis of integrated M&A strategies, the information needs are enormous (Wilbanks 2006). Especially high priorities seem to be better fine-grained climate change forecasting for both developed and developing countries, improved information about the costs and benefits of M&A actions, better information about interactions between M&A actions and other sustainable development strategies, and an improved understanding of uncertainties that surround these sorts of issues.
- *Needs for capacities* Effective integration of M&A requires two kinds of improved capacities. One need is for improved conceptual and methodological tools and improved capacities in the research communities. Another is related to the fact that effective integration also requires contributions from a wide variety of interested and potentially affected parties, many of whom do now have capacities to participate in interactive discourses with analysts. It is especially important to consider needs for capacity building of both kinds in developing countries and at relatively local scales.
- *Needs for policy-making* Integrated M&A analysis is empty unless it is connected with actions. Important initial needs include perspectives on climate policy that incorporate both M&A alternatives and structures for policy-making that are linked with analytical support. In the long run, integration of M&A to reduce risks and costs of climate change depends on effective policy-making and implementation, informed by good advice about what is likely to work and reinforced by success experiences where the advice is validated by experience. It is here, in the end, where good analysis and good action are demonstrably valuable to a society at large, while weak analysis and weak action can be demonstrably harmful.

Especially high-priority topics for future research include (Wilbanks, *et al.* 2003)

- assessing regional-scale impacts of climate variability and change;
- incorporating temperature extremes, precipitation change, sea-level rise, and severe climate events in forecasts of climate-related damages and impacts;
- characterizing mitigation and adaptation strategies, with special focus on costs/benefits and potentials/limitations of adaptation

pathways, and northern and southern regions, including attention to historical experience with adaptive behaviour;
- improving the understanding of how M&A actions are related, in some cases working at cross-purposes but in other cases offering complementarities and synergies (Wilbanks and Sathaye 2007);
- developing a new class of metrics for comparative assessment of M&A benefits (to address many non-market impacts and intergenerational issues more effectively);
- experimenting to a greater extent with different approaches to modelling integration, such as gaming and simulation modelling as well as dynamic optimization, because any single modelling approach misses some dimensions of how decisions are actually made;
- evaluating the respective roles and interactions between autonomous response behaviour and joint/collective behaviour (for example, independent decisions by individuals, reflecting their own understandings of costs and benefits versus decisions by communities and nations to act on behalf of group interests);
- examining the implications of the sequential resolution of uncertainty through learning (for example, how might reduced uncertainties about either climate change impacts or costs and benefits of response pathways affect our views of desirable response portfolios?); and
- improving concepts and tools for incorporating impacts of abrupt change as well as gradual change.

Work is also needed to further understand the implications of a wide range of uncertainties in projecting the magnitude and nature of climate change damages and the costs and performance of M&A measures. Arguments can be made for and against stressing either adaptation or mitigation under uncertainty, depending on the nature of the uncertainty, its rate of sequential resolution, and likely potentials and limitations of adaptation to cope with extremely adverse consequences.

Integrating climate change M&A is a relatively new climate change response issue, and research agendas are only beginning to appear. The highest priorities appear to be the following.

- Much better understanding of potentials, costs and benefits, and possible limits of M&A actions in particular contexts over a range of time scales in order to compare relative payoffs.
- Much better understanding of interactions between particular M&A actions in order to assess mutual co-benefits as contrasted with possible conflicts between actions.

Addressing issues in technology transfer

Research is also needed to accelerate both appropriate technology development and technology transfer. This is a special case of the more general challenge for sustainable development of improving the understanding of how to integrate knowledge into action for adaptive management and social learning (NAS 1999). Too often, the technology side of climate change responses is considered a technology development challenge as distinguished from a research challenge, when in fact prospects for a technology to penetrate markets and contribute to global problem-solving depend fundamentally on understanding how technology transfer works in a world where many conditions are going to be different from the past.

Some examples of important fundamental research priorities are listed briefly under Climate Change Mitigation R&D above, especially the item regarding research to improve basic foundations for knowledge. Other research needs that are somewhat more applied are as discussed below.

Research to accelerate appropriate technology development and use

Research needs include both accelerating improvements in the technologies available to reduce GHG emissions, either enhancing characteristics of current technologies or introducing new technologies that are superior to those currently available, and accelerating the rate of moving such better technologies into use.

Inducements for technological innovation

Although some experts have argued that it many be possible to meet technology needs for GHG emission stabilization by pushing the limits of all known technology options (for example, Pacala and Socolow 2004), most experts are pessimistic about the prospects for stabilization while at the same time, increasing the supply of energy services to the global population by a factor of three or more to enable its sustainable development (for example, Hoffert, *et al.* 2002). It seems more likely that we need not only incremental improvements in the performance of known technologies but also transformational new technologies that are far better at delivering affordable energy services with acceptable environmental effects than what we know now. If so, a significant R&D challenge is to understand how to accelerate energy technology breakthroughs. Options worth studying include the following (Wilbanks 2008b).

- *Realizing the potential of market forces* There is abundant evidence that market signals such as prices provide incentives for technology improvements, although effects on the rate of transformational change are not as well understood as on incremental technology improvement.
- *Investing in focused technology development programmes* There is also abundant evidence that large-scale technology R&D programme investments, such as the Apollo Program in the US, can lead to rapid technology improvements. Such programmes are generally highly mission-focused, but transformational changes often emerge as enabling technologies in order to meet mission requirements.
- *Enhancing linkages between basic research and applied technology development* In many cases, the critical path for transformational change is in the underlying basic science, as illustrated by Figure 1. It is at least possible that relating agendas and support for basic science needs for transformational technology development could accelerate transformations.
- *Stimulating and freely exchanging knowledge* There is a large body of research literature that suggests that the provision of knowledge can be a stronger driver of innovations than economic incentives alone. Suggested approaches include commons-like non-market information networks that encourage individual initiatives[1] and also act as platforms for person-to-person collaboration (Wilbanks 2008b).

Improving technology compatibilities with diverse market conditions Another R&D need is to recognize that all technology markets are aggregations of niche markets, not enormous collections of average

[1] For example, Science Commons: http://sciencecommons.org

conditions. Even in such industrialized countries as the US, a history of underestimating markets for such niche alternatives as hybrid vehicles and 'green power' has slowed down market penetration of innovations that are important responses to concerns about climate change. In many developing countries, the mismatch between technologies developed for industrialized country markets and local conditions is often quite dramatic. From household appliances to highway vehicles, product markets are potentially immense, especially as incomes rise, but these localities have limited influence over global technology development agendas. Meanwhile, their contexts are often quite different: a need for smaller scales, a lower threshold for affordability, a need for greater robustness, a more limited capacity to accept risks, often more limited infrastructures for technology support (for example, maintenance infrastructures), and a greater range of variation in what makes sense. In most parts of the world, in fact, a perspective that places a very high value on 'economies of scale' in technology production, associated with 'technological learning', needs to give equal emphasis to the fact that 'one size does not fit all'. Clarifying such balances between

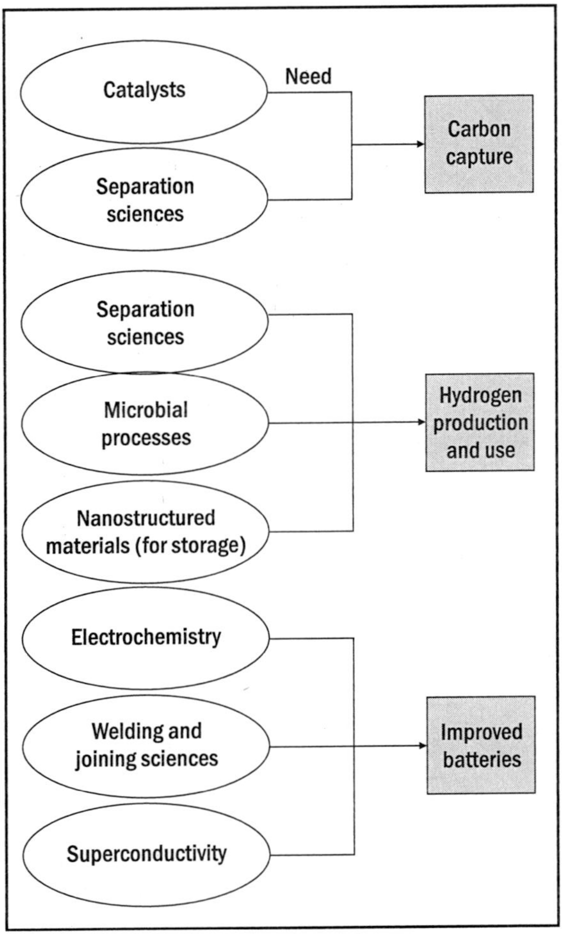

Figure 1 Linkages between basic research and applied energy R&D: focusing inward on the applied need

and among regions – and how to think about them and represent them in technology development and deployment – is a high priority for research.

Filling gaps in technology portfolios

One consequence of the limited clout of the developing world as technology R&D agendas are developed is that some technologies that would be widely useful in the developing world are under-represented in those agendas. The classic example is probably solar cooling. A great many developing countries need energy services for cooling more than for heating, at least for buildings, and have abundant solar energy resources. On the other hand, technology R&D agendas for renewable energy alternatives tend to be dominated by higher-latitude industrialized countries where cooling is not such a dominant issues; and solar energy has tended to be developed for electricity production and heating, not for cooling. Other examples include smaller-scale, simpler, robust technology alternatives (see above) and, for rapidly growing economies, implementable designs for energy and resource-efficient industrial and/or commercial complexes.

Research to accelerate appropriate technology transfer

Besides improving technologies for responding to climate change concerns, it is vitally important to improve the capability to move better technologies into the marketplace more quickly than usual; and research is needed to inform such strategies and policies.

Improving system linkages between technology research, technology development, and technology deployment

In far too many cases, the problem is not that knowledge and/or technology alternatives are not available, it is that the alternatives are not being used: either they are not known by parties who would use them or there are barriers in the way of moving them into use (NAS 1999) For example, it is common in developing countries for R&D centres to be better-connected with international R&D counterparts than with their own country's industrial firms.

A number of efforts are under way to try to find ways to bridge the gap more effectively. For example, in April 2008 the US National Academy of Sciences hosted the 'Sackler Symposium on Linking

Knowledge with Action for Sustainable Development[2], including attention to research-based knowledge on informing R&D priorities through dialogues between decision-makers and scientists, integrating knowledge from tradition, practice, and research, bridging the 'valley of death' (building end-to-end systems for linking research, development, and deployment), designing adaptive systems to link knowledge with actions for sustainability, and challenges related to ownership, governance, and cooperation.

One approach is to embrace the emerging power of the Internet to encourage relatively unconstrained free exchanges of information between providers and users, leapfrogging over traditional structures through which science and technology-related information normally trickles down – for instance by creating 'information commons' where information involving some rights to be reserved can be exchanged freely among parties who agree to certain common protocols.[3]

Improving the receptivity of markets to innovative technologies

A fundamental research challenge is to improve the understanding of practices and policy instruments that will foster an economic environment that promotes accelerated market acceptance of improved technologies. For example, a workshop in August 2008 considered a wide range of aspects of 'The Economics of Technologies to Combat Global Warming', from basic science and engineering challenges to theories of induced innovation and other economic issues related to policy instruments, technology diffusion, and modelling technological change (Nakicenovic and Nordhaus 2008), and a second workshop followed in August 2009. Papers from this series of workshops are expected to be submitted for publication in 2010. Challenges include understanding the complex interplay between market processes, changes in technology characteristics (and the driving forces behind them), institutional roles in technological change, and policy interventions to guide and accelerate technological change.

Smoothening energy transitions

Research focused on improving market receptivity is one part of the broader agenda of research to improve the understanding of

---

[2] http://www.nasonline.org/site/PageServer?pagename=SACKLER_sustainable_development

[3] Science Commons: http://sciencecommons.org

energy transitions, that is, how we shift from one global energy system to another. Other important issues include (Greene 1994): what determines the rate at which a transition proceeds? How can a transition be facilitated? Under what conditions will it be sustainable? What are the costs and benefits of one trajectory of technology/policy change when compared with other alternatives? When should investments in the new infrastructures be made? What new kinds of new institutional structures, expertise, and training are likely to be needed? Who plays which roles? How generic is the process likely to be, in a context where different nations and regions are embedded in different economic, political, and cultural traditions?

Though these needs are considerable, however, in a field which is often overlooked in climate change research support, the most important priorities for research to accelerate technological change appear to be as follows.

- Improving the understanding of accelerating transformational technological change in the next decade or two.
- Improving the understanding of strategies and mechanisms to move innovative technologies offering significant climate change benefits into widespread use in economic markets worldwide within a decade rather than over a quarter of a decade.

Conclusions

The central challenge is to significantly improve the knowledge base for responding to climate change – mitigation, adaptation, combining mitigation and adaptation, and accelerating technology development and use – while at the same time moving aggressively to act based on current knowledge.

A metaphor used widely in the early years of the US Climate Change Science Program is that, where the knowledge to guide climate change policy-making and action is concerned, the glass is half empty or half full, depending on one's perspective—and both are true. The glass is half full in the sense that a great deal is already known to guide actions taken now, as long as those actions are combined with strategies to learn from early experience. We can begin to move climate change responses in the right directions now. The glass is half empty in the sense that current knowledge is inadequate because over the longer run, we can do much better if we know more, and we can know more if we focus R&D

on the right questions. The question is not whether to study or to act. The imperative is to do both, in parallel and in interaction with each other, in order to maximize our chances to respond to climate change in ways that are effective, affordable, and equitable—one of the great challenges to science and policy in this century.

References

CCSP SAP 4.5. 2007
Effects of climate change on energy production and use in the United States, edited by T J Wilbanks, *et al.*
Washington, DC: US Department of Energy

CCSP SAP 4.6. 2008
Human settlements
In *Analyses of the Effects of Global Change on Human Health and Welfare and Human Systems*, Chapter 3, edited by T J Wilbanks, *et al.*
Washington, DC: US Environmental Protection Agency

CCTP (Climate Change Technology Program). 2006
US Climate Change Technology Program Strategic Plan
Washington, DC: US Department of Energy
[DOE/PI-0005]

DoE (Department of Energy). 2008
Workshop on Science Challenges and Future Directions for Integrated Assessment Research, November 13–14
US DoE: Integrated Assessment Research Program, Office of Science

Greene D. 2004
Energy transitions: planning and implementation.
Oak Ridge Center for. Advanced Studies. Oak Ridge, TN, USA
[Working paper, March]

Hoffert M. *et al.* 2002
Advanced technology paths to global climate stability: energy for a greenhouse planet
Science **298:** 981–987

Interlaboratory Working Group. 2000
Scenarios for a clean energy future
Prepared by Lawrence Berkeley National Laboratory (LBNL-44029) and Oak Ridge National Laboratory (ORNL/CON-476) for the US Department of Energy

IPCC (Intergovernmental Panel on Climate Change). 2007a
Climate Change 2007: impacts, adaptation, and vulnerability
Working Group II Contribution to the Fourth Assessment Report of the IPCC, Cambridge

IPCC (Intergovernmental Panel on Climate Change). 2007b
Climate Change 2007: mitigation
Working Group III Contribution to the Fourth Assessment Report of the IPCC, Cambridge

IPCC (Intergovernmental Panel on Climate Change). 2007c
Technical Summary
Working Group III Contribution to the Fourth Assessment Report of the IPCC, Cambridge

IPCC (Intergovernmental Panel on Climate Change). 2008
Towards New Scenarios for Analysis of Emissions, Climate Change, Impacts, and Response Strategies, edited by Richard Moss, *et al.*
Geneva: IPCC
[Expert Meeting Report]

Nakicenovic N and Nordhaus W. 2008
The Economics of Technologies to Combat Global Warming
Snowmass, CO, USA: Preliminary report on a workshop at the Energy Modeling Forum, August 4–5

NAS (National Academy Press). 1999
Our Common Journey: a transition toward sustainability
Washington, DC: NAS

NRC (National Research Council). 2008
Fundamental research priorities to improve the understanding of human dimensions of climate change
In *Evaluating Progress of the US Climate Change Science Program: methods and preliminary results*, Appendix B, edited by P Stern and T Wilbanks
Washington, DC: NRC

Pacala S and Socolow R. 2004
Stabilization wedges: solving the climate problem for the next 50 years with current technologies
Science 305: 968–72

Raupach M. 2007
Global and regional drivers of accelerating CO_2 emissions
Proceedings of the National Academies of Science **104:** 10288–10293

Tol R. 2002
Estimates of the damage costs of climate change
Environmental Resource Economics. **21:** 47–73, 135–160

Wilbanks T. 2007
Adaptation to climate change: how R&D can contribute
Laboratory Energy R&D Working Group, US Department of Energy

Wilbanks T. 2008a
The clean energy dilemma in Asia: is there a way out?
Eurasian Geography and Economics **49/1** (2008): 379–390

Wilbanks T. 2008b
Inducing Tranformational Energy Technological Change: some non-traditional perspectives.
Snowmass: Energy Modeling Forum

Wilbanks T, *et al.* 2003.
Possible responses to global climate change: integrating mitigation and adaptation.
Environment **45/5** (June 2003): 28–38

Wilbanks T, *et al.* 2007
Toward an integrated analysis of mitigation and adaptation: some preliminary findings
In *Challenges in Integrating Mitigation and Adaptation as Responses to Climate Change*, co-edited by T Wilbanks, R Klein, and J Sathaye
Mitigation and Adaptation Strategies for Global Change **12/5 (June 2007):** 713–25 (special issue)

Wilbanks T and Sathaye J. 2007
Integrating mitigation and adaptation as responses to climate change: a synthesis
In *Challenges in Integrating Mitigation and Adaptation as Responses to Climate Change*, co-edited by T Wilbanks, R Klein, and J Sathaye
Mitigation and Adaptation Strategies for Global Change **12/5 (June 2007):** 957–962 (special issue)

5

Climate change: a problem and an opportunity

Bjorn Stigson*

The World Business Council for Sustainable Development (WBCSD) and its member companies are working on climate change as the key issue for this century. There is need to establish an equitable, clear, long-term approach to dealing with the challenge in an integrated, multi-faceted, and economically sustainable way. The scale of change will mean nothing short of a new industrial revolution.

Energy is essential for economic growth and social development. Over the coming decades, there will be a surge in energy demand as developing economies grow. Fossil fuels will provide most of the planet's energy at least up to the middle of this century, leading to a considerable rise in greenhouse gas (GHG) emissions, especially carbon dioxide (CO_2). This trend must be reversed by decarbonizing the world's energy mix and using resources more efficiently to stabilize and then reduce GHG emissions.

Views in the international climate policy community indicate that the recent financial crisis and economic slowdown will make it more difficult to reach agreement on a post-Kyoto framework. However, Sir Nicholas Stern begs to differ: 'The pluses outweigh the minuses... [the current financial crisis was] 10, 15, 20 years in the making – it was possible to understand how it was building up. Climate change is very similar. The current crisis is showing us the consequences of ignoring our long-term actions.'

---

* President, World Business Council for Sustainable Development (WBCSD), Geneva

The Chairman of the WBCSD, Global CEO of PriceWaterhouse Coopers, Mr Samuel A DiPiazza Jr, concurs: 'The need for responsible business engagement in the big issues of our time has never been more urgent. The current financial crisis is the result of short-term and unsustainable business models. Our experience is that companies that have put sustainability issues at the heart of their business strategy offer some valuable lessons for those negatively impacted by the crisis. Different industries at different times have had to understand how sustainability issues such as constraints on carbon, water and ecosystems, or social impacts affect their profitability.'

The companies in our membership are innovative and well-managed, partly because of their commitment to sustainable development. They are also frontrunners in understanding the business benefits and opportunities derived from innovative energy and environmental approaches. We must now develop business models for growth driven by a low-carbon economy.

The scale of change required

In September 2008, Yvo de Boer, executive secretary of the UN Framework Convention on Climate Change (UNFCCC), told the first Africa Carbon Forum, Dakar, in his opening speech: 'Combating climate change will take political will, and it will take a great deal of investment...What the world needs is a global, low-emissions economic development plan that makes climate-friendly economic growth globally viable.'

The International Energy Agency's (IEA's) *World Energy Outlook 2008* reference scenario, which assumes no new government policies, sees world primary energy demand growing by 1.6% per year on average between 2006 and 2030 – an increase of 45%. This is slower than projected last year, mainly due to the impact of the economic slowdown, prospects for higher energy prices, and some new policy initiatives. China and India account for over half of incremental energy demand up to 2030, while the Middle East emerges as a major new demand centre.

This means that three-quarters of the increase in carbon dioxide emissions will come from China, India, and the Middle East, and 97% from non-OECD countries as a whole. If populations in these economies are to achieve the levels of development and standards of living to which they aspire, and if we are slow to improve resource efficiency and limit climate change impacts, the use of low-carbon energy must account for at least 36% of the global primary energy mix by 2030, according to the IEA.

We have identified six key areas where large changes are needed to help stabilize GHG emissions.

1 Power generation
2 Industry and manufacturing
3 Mobility
4 Buildings
5 Consumer choices
6 Land-use changes.

We have posited that by 2050, the power sector deliver low-GHG electricity from a variety of sources, and mobility is characterized by high-efficiency vehicles, new fuel technologies, and a balance between private and mass transportation.

Turning such assumptions into realities will require the following.

- Rapidly deploying existing low-GHG, energy-efficient technologies encouraged by a range of supporting government policies;
- Developing new, breakthrough technologies over the next two decades;
- Increasing consumer awareness and understanding of energy issues and GHG emissions, leading to greater demand for energy efficiency at all levels.

A singular focus will be insufficient, although energy efficiency will play the more dominant role during the 2010–20 period.

For a country such as the US, it amounts to a long-term improvement in energy efficiency of about +2.5% per year (vs. 1.5% per year now) and a decarbonization rate (CO_2/energy) of some 1% per year until 2020 or 2025, then more than doubling in the period after that as technologies such as carbon capture and storage (CCS) can be deployed at scale.

Developing country commitment will initially require much greater use of market mechanisms, for example, as a bridge to setting emissions-intensity-based targets within the economy or introducing technology standards.

The opportunities

International cooperation: a global framework built on national approaches

Although there is some debate about whether peak oil production has been reached, with some arguing that we have already started the downward trend towards depletion, it is widely acknowledged that fossil fuels, particularly coal, will be the primary source of energy for the foreseeable future. This is especially true in India, China, the US, and South Africa, where coal is abundant. At present, some 40% of electricity generation is coal-fired. Yet, burning fossil fuels is responsible for GHG emissions and pollution, and is a major contributor to climate change. So, the big question is how can we meet the energy needs of a growing developing global population while not further exacerbating climate change with all of its consequences?

The answer will lie in substantially improved efficiency in energy use, in developing alternative, renewable, non-polluting sources of energy (wind, wave, solar, nuclear), and on making existing sources of energy cleaner and more efficient (clean coal, CCS, and so on). This will require the participation of both developed and developing nations.

As purveyors of technology and technological development, WBCSD members have been investigating and developing innovative technologies to make this possible. However, business will not be able to do this in isolation. In order to make this happen, business requires comprehensive policy frameworks to guide its research and development activities and set goalposts for the objectives it is seeking to achieve. This will require the articulation of firm pathways to emissions reductions, and increased commitment and greater financial investment on part of governments.

Energy use and infrastructure

Energy policy is set at the national level against a backdrop of prevailing financial, security and environmental signals. A climate change policy framework must recognize the sovereign nature of energy policy decisions but, at the same time, provide clarity and a context within which such decisions are made. The framework should provide the tools to optimize GHG emissions management and direct capital towards low- and zero-GHG emissions technologies.

Elements that could collectively form the basis of a flexible and diverse framework to support GHG emissions reduction efforts include the following.

- Establishing by 2010 a quantifiable, long-term (to 2050) emissions pathway for the management of global GHG emissions and actively disseminating its meaning through all levels of society.
- Encouraging the development and deployment of leading-edge technologies through partnerships and incentives and an approach to mitigate long-term market risk and deliver secure benefits for large-scale, low-carbon, new technology projects.
- Including ideas and lessons learned from current approaches and instruments that are being used and continue to evolve at international, regional and domestic levels.
- Modifying the existing international framework so that it builds progressively (bottom up) from local, national, regional or sector programmes that contribute to the quantifiable long-term international pathway and catalysing the implementation of such programmes.
- Allowing industry sector participation across multiple facilities or technology platforms at the national level and across national boundaries, and enhancing GHG project mechanisms to allow them to cater for sector projects.
- Progressively including all countries – both developing and developed.

Land-use change and forestry

While deforestation and forest degradation produce about 20% of the world's GHG emissions, the forestry sector also has the unique ability to simultaneously reduce emissions, capture and store carbon, and lessen the vulnerability of people and ecosystems to climate change.

Forest-related mitigation measures in the short to medium term are among the easiest to implement and the most cost effective, have very low opportunity costs and can make an immediate and direct contribution to sustainable development and rural livelihoods.

Measures such as sustainable forest management, forest conservation, reforestation, forest restoration, afforestation, wood-based bioenergy generation, and the use of sustainably produced wood products as substitutes for emissions-intensive materials, therefore, should all be considered as part of the global approach to climate change mitigation and adaptation.

However, forest-based climate change responses should be designed as a complement, rather than an alternative, to strategies for reducing fossil fuel emissions. A strong focus on forests should diminish neither the need for clear emissions limits nor the climate change mitigation responsibilities of the major emitting countries.

Experience within the forest community has shown that, in the long run, a narrow focus on a single commodity – such as carbon – at the expense of wider forest values is unlikely to succeed. Moreover, a piecemeal approach to forests would risk undermining the social, environmental, and economic resilience of rural communities and could directly reduce their ability to adapt to climate change. To be effective, forest-based mitigation and adaptation measures must deliver sustainable development, build resilience in rural communities, and fully involve forest-dependent people, civil society, the forest workforce, and the private sector.

Five principles, described below, should guide forest-related post-2012 arrangements on climate change.

1. Ensure that forest-related climate change options support sustainable development in both forest-rich and forest-poor countries.
2. Tackle the drivers of deforestation that lie outside the forest sector.
3. Support transparent, inclusive, and accountable forest governance.
4. Encourage local processes to clarify and strengthen tenure, property, and carbon rights.
5. Provide substantial additional funding to build the capacity to put the above principles into practice.

The WBCSD recognizes that climate change mitigation and adaptation measures will only achieve lasting results if they are adapted to conditions on the ground and help meet the needs of local people. By providing adaptability and supporting livelihoods, sustainable forest management can ensure healthy and productive forests, underpin robust rural livelihoods, and deliver a wide range of products and ecosystem services that societies demand. It can also be an economically, environmentally, and socially effective way of addressing climate change globally.

Aligning a new international climate change framework

The principal existing international framework, the Kyoto Protocol, is broadly a 'top-down' approach. By contrast, energy production and use patterns develop largely 'bottom up' from local, national and regional policies, coupled with the availability and security of energy resources. Aligning a new international climate change framework with existing approaches to energy access and security issues would offer greater scope for encompassing the large-scale changes needed in the energy system.

A quantifiable long-term pathway, established collectively by governments, scientific and technological organizations, business, and civil society, would assist in reducing current levels of uncertainty. The long-term pathway becomes a point of reference for the development of national energy and climate policy and raises consumer awareness of the need for GHG emissions management.

All technologies face barriers of some sort. GHG emissions management technologies face additional hurdles posed by the uncertainty over long-term political action to address climate change. A diverse approach will allow business to better manage this additional risk and further encourage technological development and deployment.

For WBCSD members, the next industrial revolution has now begun in earnest. The WBCSD is concentrating on advocacy to establish a post-2012 framework, concentrating on input to international energy and climate policy-making and developing content – and technology – to support these.

Technology development, deployment, and transfer

Reducing the energy intensity of the global economy and the GHG intensity of energy will require major breakthroughs in energy efficiency, renewables, next generation nuclear, clean coal, CCS, and mobility. Such breakthroughs must be developed and then deployed.

Both require effective government support mechanisms. If the development phase is too costly for the private sector to sustain alone, technologies will never be able to enter the market and achieve full commercialization. Governments need to support new technologies during these critical early stages, establishing clear priorities for research, development and demonstration (RD&D).

Each technology is different and thus requires different enabling policies. For technologies that are already cost effective, policies need to focus on barriers to investment or restrictions to market entry. Emergent technologies, ready for rapid deployment, need government supports such as subsidies, tax credits, capital allowances and depreciation schemes, certificates, and tariffs. Governments must take account of the full development and deployment cycles for their national situations (for example, wind and hydro).

Long-range breakthrough technologies such as clean coal and next generation nuclear need to be guided the whole way through research, development, demonstration and deployment. Governments can do this through research partnerships with business or bilateral or multilateral corporations. Changing markets and consumer attitudes to speed deployment also requires appropriate policies. For example, many low- and zero-GHG emission technologies (such as CCS) will not succeed without carbon pricing and direct forms of government assistance Otherwise, there is no genuine business case to invest.

In order to give business the incentive to invest in technology development and deployment, governments must first decide a long-term global GHG emissions pathway in terms of carbon equivalent emissions. Such a pathway should be in place by 2010 and revisited periodically, taking into account the development of climate impact science and the rate of technology deployment. Depending on the rate of technology development and the trajectory targeted, deployment would depend on ensuring that all new projects use the new technology, together with an appropriate balance between early removal of existing capital stock and existing capital stock remaining to live out its life.

The introduction of a long-term pathway should encourage new technology development and the implementation of medium- to large-scale demonstration projects. But projects that are purely climate driven (such as CCS) simply will not happen unless there appears to be a future market or demand for such a service or technology. Two approaches could be used to overcome such risk.

1 Direct incentives for technology programmes, such as cooperative clean development networks at the international level or policies that encourage the development of new technologies through RD&D assistance, capital allowances for new low or zero-GHG infrastructure and early take-up incentives and consumer education

programmes designed to bring new products and services to the market more rapidly (for example, GHG labelling or consumer websites), at national levels.

2 Managing the long-term regulatory risk of a new technology project through the use of multilateral financing mechanisms or a mechanism that underwrites the long-term validity of the associated GHG reduction units.

Signatories to the long-term international objective would be expected to develop national programs to manage GHG emissions or encourage industry sector programmes in alignment with the agreed global emissions pathway. Cross-border industry sector programmes could also be developed.

Cooperation on technology development could lead to international agreement on technology codes and standards for industry and manufacturing, buildings, mobility and consumer products, thus hastening the adoption of new technologies globally.

Energy efficiency and business

In our first book, *Changing Course*, the WBCSD highlighted the business opportunities presented by the challenge of sustainable development and coined the term 'eco-efficiency'.

We define eco-efficiency as achieving the delivery of competitively priced goods and services that satisfy human needs and bring quality of life, while progressively reducing ecological impacts and resource intensity throughout the life cycle to a level at least in line with the earth's estimated carrying capacity. This is commonly summarized as 'creating more value with less impact'.

Eco-efficiency became an influential concept with business and governments because it links energy efficiency, environmental performance, and competitiveness. Eco-efficiency is primarily a business concept because becoming more efficient makes good business sense. In the case of energy efficiency, it calls for businesses to achieve more value from lower energy input and applies throughout the company – to marketing and product development as much as to manufacturing or distribution. Three important lessons from our eco-efficiency work also hold true for energy efficiency.

1 Companies can re-engineer their processes to reduce the consumption of resources and pollution while saving costs.

2 By cooperating with other companies, many businesses have found creative ways of reusing and revalorizing their by-products.
3 Redesigning products also helps to become more eco-efficient as well as to find new ways of meeting customer needs.

Energy efficiency

Energy efficiency is widely portrayed as the most cost-effective way to mitigate climate change. According to the IEA, energy efficiency accounts for more than half the potential to bring CO_2-emissions back to today's levels by 2050 or reduce them further. The bulk of energy efficiency measures will have to happen on the demand side (in industry, manufacturing, commercial and residential buildings, and transport), and changing consumers' behaviour is key.

The business case for energy efficiency is clear: It reduces energy costs, diminishes vulnerability to increasing energy prices, and reduces CO_2 emissions. The benefits of energy-efficiency measures generally outweigh the costs. However, despite a simple cost-benefit approach, energy efficiency faces barriers when it comes to implementation. Energy efficiency cannot be achieved by business alone. It needs governments to create enabling frameworks and society to change consumption patterns in order to help companies provide more efficient products and services.

Energy efficiency: facts and trends

- More than 50% of the GHG emissions reduction potential lies in energy-efficiency measures: end use fuel efficiency (24%), end use electricity efficiency (12%), end use fuel switching (11%), and power generation efficiency and fuel switching (7%).
- End use efficiency can contribute almost one-third of the necessary emissions reductions while generating macro-economic benefits (that is, costs < benefits)
- Buildings account for up to 40% of primary energy use in most countries, and consumption is rising. The IEA estimates that energy demand for buildings will stimulate about half of energy supply investments to 2030.
- In IEA countries, between 1990 and 2004, energy efficiency improvements led to energy and CO_2 savings of 14% respectively, which translates into at least $70 billion of fuel and electricity cost savings in 2004.

Energy efficiency is one of the most obvious possibilities for mitigation actions in developed and developing countries and it is indirectly included in the decisions adopted by the 'Bali Action Plan'.

International climate negotiations face the double challenge of creating the right incentives for a massive shift to cleaner energy technologies and a more productive use of energy while at the same time finding a fair 'global burden sharing' mechanism that neither compromises the growth trajectories of today's developing countries and emerging economies nor the living standards in today's industrialized countries.

Energy efficiency is the low-hanging fruit for energy savings in both developed and developing world and should be accelerated. Based on its member's business experience, the WBCSD has identified the following measures that will help achieve further energy efficiency gains.

- Identify and eliminate perverse subsidies.
- Internalize environmental cost.
- Develop and implement economic instruments such as emissions trading.
- Promote voluntary initiatives and negotiated agreements.
- Develop and implement energy-efficiency standards for industry applications, housing, and mobility.
- Boost best practice sharing and energy-efficiency education.

Appropriate measures must be created to get consumers to focus on energy efficiency topics, because today, most consumers do not really want to deal proactively with energy questions. Incentives, product information, and leadership need to be coordinated to change end-consumer behaviour.

Among other things, companies can make public commitments to speed up the process and gain support from employees, suppliers, partners and customers, translate organizational energy conservation commitments into targets and accountabilities for specific business units/sites/processes, collaborate to reach widespread adoption of breakthrough technologies, and develop and commercialize products and services that help consumers reduce their life-cycle energy use (for example, informing users about energy use patterns, helping them use products and services at maximum efficiency).

Companies can also disseminate information systematically and repeatedly to raise public awareness about opportunities to adopt energy-saving measures that can result in economic benefits.

Governments, for their part, could seize the opportunity to learn from successful regulatory standards and to seek to harmonize them worldwide. Particular emphasis should be given to systems that gradually crowd out the most inefficient technologies. They could also provide targeted purchase incentives such as tax rebates, targeted subsidies, and preferential loans to help companies jump the hurdle of high up-front investment costs for more efficient technologies.

Additionally, governments could develop financial mechanisms that reduce the initial capital burden related to the purchase of efficient end-use technologies (such as tax credits, mortgage discounts, rebates, preferential loans) and remove split incentives and other financial barriers. They could also use a holistic approach to building energy systems to encourage interdependence between different stakeholders and find a balance between energy security and poverty alleviation while abolishing subsidies on non-renewable energy sources that do not serve an equity purpose, thereby distorting incentives for efficient energy use.

Finance

The need to address climate change while facilitating continued economic growth and social progress is one of the key challenges facing world leaders today. Energy is critical to continued economic growth and according to the IEA, population growth and increasing industrialization will drive demand for energy upwards by more than 50% between now and 2030.

Fossil fuels are expected to be a major contributor in meeting these future energy demands. This will lead to a substantial increase in GHG emissions unless cleaner technologies, such as carbon capture and storage (CCS), renewables, and nuclear power, emerge that are commercially competitive and can be implemented at scale. The demand for energy will rise most rapidly in developing countries as they develop energy services to drive economic growth and social progress. Without convenient and affordable alternatives, these countries are likely to follow a high-carbon pathway, similar to that of the developed world.

The WBCSD has long explored the highly political issues concerning the challenge of meeting energy needs without causing irreversible damage to the earth's climate.

The solutions lie in creating framework conditions with the right incentives to bring about a large-scale technological shift toward a lower-carbon and more energy-efficient economy that also delivers affordable energy solutions for the 2.4 billion people who are currently without basic energy services. This shift relies on scaling up investment flows into the development and deployment of lower carbon technologies, as well as adapting behaviours and lifestyles to favour these technologies across the developed and developing world.

The private sector is a major source of innovation, capital and capacity that, given the right framework, can deliver a low-carbon global economy. For governments to facilitate the release of private sector resources, they need to understand how capital markets and corporate investment strategies can be incentivized to deliver results consistent with sought-after goals on carbon mitigation and improved access to energy.

Many companies are already using entrepreneurial innovation to adapt their business models and develop strategic approaches to reduce GHG emissions. Yet business can only go so far on its own. Before a company invests, it studies the future. If governments do not establish equitable policies that signal a future in which energy efficiency and a lower GHG energy mix are going to be valuable to companies and society, then business will have trouble justifying investments in such innovations. Public investment and RD&D-friendly policy frameworks will foster business commitment in the early stages of development cycles.

In its 2008 report entitled *A Warming Investment Climate*, Goldman Sachs suggested that, 'while equity markets are focused on more immediate concerns, structural trends will continue to reshape competitive landscapes in global industries through this and future economic cycles. Moving to a low-carbon economy will entail significant behavioral change across society...the drivers of competitive advantage will be shaped across all industries and regions.'

Goldman Sachs also indicated that they saw, 'few indications that economic slowdown will diminish interest in emerging carbon markets.' Other economic commentators such as Sir Nicholas Stern have indicated that as a direct result of the financial crisis, the carbon market could flourish as a much needed driver of growth.

To be a force for change, governments must help to reduce uncertainty for investments in higher cost, lower emitting systems. Government support will remove barriers and uncertainty, facilitate the establishment of efficient markets, and drive technological innovation and investment.

- International political frameworks must be aligned with the long-range business investment cycle so that investments in CO_2 abatement technologies can be justified commercially.
- Many technology projects require government policies on issues such as RD&D, risk management, and large demonstration projects.
- Utilizing a range of possible instruments, price signals should be created to promote innovative product and technological design. Policies that promote GHG emissions reductions will send the required signals to capital markets.
- Strong financial commitments by multilaterals will encourage development and transfer of leapfrog technologies to developing countries. Good governance must accompany additional financing and technical support for energy and technology markets.
- Policies must include education programmes attracting consumers towards low-GHG products and services.

No single action will meet these challenges while sustaining economic growth. An integrated, multidimensional portfolio of technologies, policies and mechanisms will be required — along with a balance of global, regional, and national approaches.

There are, however, quite different mental models and perceptions among governments and policy-makers of how and why a business might choose to invest in a particular project and/or country. This can result in inefficient policies that hinder rather than support the involvement of the private sector.

Energy is a key driver of economic growth and social progress. It is essential to fuelling industry, powering infrastructure, connecting goods, people and services to markets, and delivering basic services such as heating, lighting, and cooking. For the billions of people without access to modern energy services to escape poverty and enter into productive economic activities, investments in energy infrastructure (on and off-grid) are needed.

Scientific evidence and economic analysis confirm the need for radical changes in the global energy system to combat climate change and ensure energy security in the future. The UNFCCC in its 2007 analysis of financial flows estimates that some $200 billion will be needed annually up to 2030 to stabilize GHG emissions at today's levels. The incremental costs of low-carbon investments in developing countries are likely to be at least $20–30 billion per year.

Today, private sector investments constitute the largest share (86%) of global investment flows and are expected to be essential to addressing climate change. A large additional flow of tens of billions of dollars will also be needed for adaptation.

Unless policies change and ways are found to facilitate investments in lower carbon technologies at all stages of their development and deployment, developing countries are expected to follow the same carbon-intensive development pathways of today's industrialized nations. This would constitute a lost opportunity of immense proportions, as the consequences of carbon- and energy-intensive investment decisions made today lock in those emissions for decades.

Raising finance is not necessarily the main problem. Building capacity to absorb technology and implementing framework conditions that direct financial flows towards the development, demonstration and deployment of commercially viable low- and zero-carbon energy technologies is the key.

Many factors come into play to encourage a business to invest in low-GHG energy solutions, not least the strategy of the company and its ability to attract capital to implement this strategy. For policy-makers to encourage business to invest, they must understand what might incentivize a business to do so. Commercial investments are almost always made in dynamic, highly differentiated, competitive environments, both internal and external to the company.

Money normally flows to where the highest and quickest returns can be made. Key questions include: How to most efficiently allocate private capital? How long before there is a positive cash flow? What risks are associated with the investment? How, where, and when to enter new markets? Project-based investments in emerging and lower GHG energy technologies are some of the more complex and risky forms of investment. They are normally highly capital intensive and play into a world where the average consumer is unwilling to pay a premium for low GHG energy services.

Each project requires a detailed evaluation of the prospective rates of return, investment and technological risks, as well as sources of competitive advantage. Companies, financial institutions, and investors almost always employ different processes and screening criteria to evaluate the investment.

The outcome of the analysis is a forecast of the investment's ability to deliver returns evaluated against a host of risks and other possible uses of that same capital. Every decision relies heavily on the expected commercial return.

Most low- and zero-GHG energy technologies will not be cost competitive at scale without some combination of investment support mechanisms, technological advances or regulatory regime improvements. An abundance of potential projects, technologies or investment opportunities will not in itself necessarily translate into the mobilization of capital flows for implementation. Many energy-efficiency projects and Clean Development Mechanism (CDM) proposals have faced this problem, yet they have great potential for reducing emissions, managing energy demand challenges, and optimizing value in the long term.

Some of the world's poorest countries may be at a further disadvantage because of limited institutional and commercial capacity, not to mention high-risk ratings that affect the ability to attract, develop, and manage substantial project-based investments.

WBCSD members recognize the societal need to significantly scale up investments in lower GHG technologies. They also see opportunities, for example, in the following.

- Establishing new markets for low-GHG technologies and services
- Gaining competitive advantage through technological innovation
- Reducing costs through energy efficiency.

However, the members also recognize that a project's ability to attract investment also depends heavily on the prospect of a commercial rate of return. A lack of certainty over policies related to carbon pricing and GHG reduction targets increases the risk of achieving a commercial return for low-GHG technology projects. While this uncertainty prevails, the bulk of potential private capital available will probably flow to traditional energy sources, or remain uncommitted until definitive policies, which underpin a pragmatic approach, begin to emerge.

International policy efforts must align with the investment cycle that can last for decades, from initial R&D through to actual deployment at scale. A broad and efficient mix of policies and programmes targeted at mitigation and adaptation and backed by supportive regulation and governance frameworks will reduce investment uncertainty and encourage business to invest for the long term.

Governments are engaging with business through instruments like the CDM and public–private technology partnerships. These are forming new areas of collaboration and new business opportunities. However, in its current form, the CDM will fail to achieve investment at scale and will not help solve the dire energy needs in regions such as sub-Saharan Africa.

Most stakeholders agree it will take a combination of financial mechanisms, including the carbon markets and official development assistance (ODA), to guarantee that the energy demands of the future are met in a way that mitigates climate change. In order for business and private capital to play its role in delivering low- and zero-GHG technologies, key considerations in the design elements of future frameworks include the following.

Governments need to create robust and integrated policy frameworks that create predictable future demand for new technologies and reward innovation and establish a clear and strong expectation of a carbon price in the near and long-term future to encourage investment. They also need to incorporate energy and climate strategies into national development plans and develop approaches that expand or aggregate projects through programmes or portfolios to standardize and streamline the transaction process and to establish stable and transparent regulatory regimes to help reduce corruption and improve country risk profiles.

All stages in the technology development cycle need to be addressed: investment in public and private energy R&D with the support of international financial institutions to help low-GHG technologies such as CCS, renewables and nuclear power through the various stages of development; ensuring the commercial viability of technologies through direct support and incentives in the demonstration phase; adopting pragmatic and inclusive approaches that create fast-track approval processes to accelerate deployment of these new technologies; and setting an example for other sectors by acting as an early adopter, and buying new, advanced technology products for government fleets and operations.

Policies are needed to enhance growth and competitiveness in developing countries through technology cooperation by establishing a competitive business-to-business framework for transactions, dismantling trade barriers affecting the diffusion of technologies to encourage investment and business participation and managing the intellectual property rights regime to balance the need to incentivize innovation and the dissemination of technologies to support investment in new technologies.

And to cap it all, there is a need to build institutional capacity to translate policies into robust and integrated development plans, as well as to support investment in small and medium enterprises (SMEs), particularly in capacity building, so they can own and/or operate small-scale energy projects in order to help ensure deployment of technologies. Influencing public behaviour and the acceptance of new technologies through awareness raising and education will also help ensure future demand for low-GHG energy services.

Markets and business

The term 'carbon market' refers to the financial mechanisms that are in place that allow emitters (companies, countries, and individuals) to control their carbon emissions and meet mandatory and voluntary emissions commitments. Carbon markets trade not in existing carbon, but in the reduction of carbon emissions into the atmosphere. Under the Kyoto Protocol, parties are allowed to use market mechanisms such as emissions trading and project-based credit trading to meet their legally binding commitments. These mechanisms fall under two categories.

1 Allowance trading, such as in cap-and-trade schemes, where a GHG absolute target ('cap') is set (mandatory or voluntary) and subsequent trading of emissions allowances between emitters to meet the emissions target occurs. These markets are mandatory or voluntary. The markets attribute a market value to emissions reductions, allowing companies that have targets to find the most cost-effective means to reduce emissions. These markets are generally established at the regional level, such as the European Union Emissions Trading Scheme (EU-ETS), or at the national level, like the Australian New South Wales Greenhouse Gas Abatement

Scheme (NSW GGAS) or the voluntary Chicago Climate Exchange (CCX) in the United States.

2 Project-based credit trading, such as the CDM and the Joint Implementation (JI) mechanism, established under the Kyoto Protocol, in which trading in emissions reduction offsets (or credits) are undertaken at sources outside a cap-and-trade programme. These projects aim to increase the sustainable development of developing countries through the transfer of clean technologies that reduce net emissions in those countries. These projects earn saleable credits (Certified Emission Reduction credits in CDM projects and Emission Reduction Units in JI projects) that can be used to meet a company's (or other party's) mandatory commitments or voluntary compliance goals.

In recent years, carbon markets have become the most visible result of early climate mitigation efforts and have spawned a market valued at $64 billion in 2007. Within the EU-ETS market, by far the largest trading market, 775 million tonnes of CO_2 changed hands in the first half of 2007, with a total value of $11.5 billion. In 2007, 597 million tonnes of primary Certified Emission Reductions (CERs) were traded, with a value of $9.42 billion.

The UNFCCC reports that the carbon market, already playing an important role in shifting private investment flows, will have to be significantly expanded to address the needs for the additional investment and financial flows that are necessary to reduce climate change.

We at the WBCSD believe a price for CO_2 must be established to drive demand for compliance instruments, such as the existing CDM, so that a truly global market develops. The market would be used to provide an outlet for project mechanism units (credits or offsets) and link together disparate national and sectoral programs. These could include, for instance, programmes to promote energy efficiency, more use of renewables, the expansion of nuclear energy, performance targets, offsets, carbon taxes or technology funds.

Over time, more countries will adopt emissions management. National and regional carbon markets must be linked together. A global 'GHG market' would allow emissions trading between nations, sectors and projects, thereby introducing flexibility into the attainment of national and sectoral objectives. To enable links to be made, similar structural elements must exist. Important considerations include the following.

- Pricing measures – Price caps and price floors could undermine linking. A price ceiling in one system would effectively flow through to other systems once met. This could further complicate system design and operation.
- Penalties – The penalties for non-compliance must be broadly similar in the linked systems. Each must require environmental make-up of any missed target in the next compliance period (that is, no buy out) and any fines should be at a similar level.
- Banking and borrowing – Provisions for banking and borrowing should be similarly structured. Limits in one system would be circumvented by trading into a system where any limit in place had not been reached.
- Monitoring and reporting – The monitoring and reporting guidelines in each trading system should follow similar protocols.
- Linkage and offset policies – Ideally, linked systems should not impose any limits on trade flow (for example, a limit on the percentage from linked systems that may be offered for compliance).
- Legal – An international set of accounting and verification principles, and common contracts. Different approaches to allocation can exist across linked systems. Grandfathering may be offered in one system versus auctioning in another. The allocation process does not directly impact trading and price behaviour.

International GHG markets would continue to play a role in a revised framework, directing energy investment capital in favour of low- and zero-GHG emissions projects. The market would be used to provide an outlet for project reduction units and link disparate national and international sector programme together. Such national and sector programmes need not be based on emissions trading (such as cap and trade). For example, GHG emissions could be decreased through programmes to promote energy efficiency, more use of renewables, the expansion of nuclear energy, performance targets, offsets, carbon taxes, regulation or technology funds.

To introduce additional flexibility in achieving national or sector objectives, these programmes may wish to link to international emissions trading markets, provided they meet certain criteria for entry (for example, the policy measure should recognize the international trajectory and represent a realistic and tangible contribution to GHG reductions). A linked programme would become eligible for an allocation of tradable GHG allowances or would be treated as a project and granted tradable GHG reduction units. The body administering national climate policy would be responsible for distribution of GHG allocation units to the industries involved in the sector within a country.

But from a long-term perspective, what is the current role and outlook of emissions trading and project mechanisms?

The trend towards emissions growth, by some estimates by more than 50% between now and 2030, must be reversed by managing emissions, decarbonizing the world's energy mix, and using resources more efficiently to stabilize and then reduce global GHG emissions. No one single approach is the 'silver bullet'. However, flexible market mechanisms such as emissions trading and project mechanisms are a key component to addressing mitigation efforts and can play a major role in putting us on a global trajectory towards achieving long-term, quantifiable management of GHG emissions by 2050.

Business action to address climate change is already happening, and the pace at which initiatives are developing is increasing. Many initiatives are in response to national emissions objectives. Some are driven voluntarily by business.

A number of different emissions management approaches are being proposed, including economy-wide 'cap-and-trade' systems, low-carbon fuel standards, national renewable energy obligations, energy-efficiency targets, carbon taxes, and technology standards. Each approach creates a (different) cost of carbon within its targeted sector or country, either explicitly through an allowance price or implicitly through the incremental cost of policy requirements.

A key objective in creating a global carbon market is to link different emissions management approaches together, thus establishing a single carbon cost and creating equitable access to the prevailing lowest cost abatement opportunities. However, such a global market is not a substitute for policies and instruments that are designed to promote strategic investments in technology development and to provide support for large-scale infrastructure investments.

The carbon market as it stands today experienced extremely rapid growth in 2007, as it more than doubled in value at $64 billion, compared to $31 billion in 2006 and $11 billion in 2005. The EU

ETS remains by far the largest and most successful emissions trading market, and accounted for nearly 80% of the entire carbon market in 2007. Project-based transactions under the CDM also increased in 2007, up 22% from 2006 to value $7.4 billion (primary CDM only).

The overall objective of the climate mitigation policy is to manage global GHG emissions on an absolute basis, that is, tonnes of CO_2 (equivalent). Within a global emissions target, it must be recognized that some nations are still developing and their emissions will grow (with the potential to decline in the future), while other nations have the opportunity to see their emissions plateau and begin declining immediately. Nevertheless, every nation able to implement significant mitigation measures needs to structure its CO_2 emissions management programme on an absolute basis for specific future periods, at least at the sector level where the mitigation opportunities lie.

Of course, some nations are not yet in a position to begin absolute emissions management at all, but actions such as emissions reduction projects could be implemented. Absolute emissions management is key to linking disparate national programmes. However, individual programmes need not be based on absolute emissions provided the national government is prepared to balance the national accounts. As such, there are two ways to cascade a national carbon budget into the economy.

- *One-for-one cascade* In this case, there is a direct relationship between the allowance allocation at industry level within the economy and the national emissions target. The government has directly cascaded (part of) its target down into the economy, such as in the EU-ETS.
- *Indirect cascade* This means that the government has established a mechanism to manage emissions within the economy that is not directly linked to the national target. For example, it may opt for a carbon tax or establish a 'renewables obligation' in the power generation sector, which even when met does not guarantee a certain emissions outcome for the sector as a whole. However, such an approach does not negate the possibility of linking with other national programmes.

Linkage occurs when one system recognizes the market instrument (for example, allowance) operating within another system and allows its use to meet the compliance objective of the first system.

For example, the EU-ETS recognizes the CDM and allows the use of CERs – the market instrument of the CDM – to meet the compliance requirement of a facility in the EU-ETS.

A unilateral approach exists when one government recognizes the instrument of another and accepts it as legal tender, with the other government either oblivious to it or choosing to ignore such recognition.

Recognition occurs frequently in monetary markets, with the US dollar widely recognized and accepted, even though it may not be the official currency of a particular nation. In emissions markets, government recognition, even if unilateral, should be encouraged for the CDM, such that at the very least, the CER remains the de facto international unit of carbon trading. Establishing different project mechanisms to support various national systems is not only expensive from a transaction perspective, but could undermine the rigour needed to ensure project validity and environmental integrity. It could also undermine the concept of a global carbon market.

A bilateral approach involves specific recognition between two parties. While a global market could be constructed through such an approach, the transaction costs will inevitably rise for participants. For example, if A recognizes B, and B also recognizes C, then in practice, allowances from C will be transferable to A. But if A does not formally recognize C, then additional pass-through trades would need to be constructed, all at additional cost.

A multilateral approach could develop either by agreeing to a standard set of rules that govern all applications of emissions management, or by the development of a global registry and common emissions management instrument that allows international trade to take place. While both require significant levels of agreement, the latter is likely to be more flexible in that individual approaches would only have to recognize the international registry and its rules.

As is the case with the CDM, projects form an essential part of the market. They encourage market participation across a broad spectrum, ranging from the poorest developing countries where economy-wide emissions management simply is not feasible or appropriate at this time, to the richest developed economies where conventional policy instruments may be impractical – for example, 'cap-and-trade' in the agricultural sector.

As discussed in Policy Directions to 2050, the project mechanism should be broadened to include sector-wide programmes and augmented with additional instruments to encompass activities such as large hydro, nuclear, CCS, land-use change and avoided deforestation.

Project mechanism reduction units also offer a convenient international instrument for de facto linking of a wide range of mitigation approaches.

A global agreement that leads to the 'ideal' situation of a global cap with all nations participating in a single carbon market is an unlikely outcome of current international processes. Rather, a stepwise approach to this goal may be the more prudent avenue to follow, through an arrangement that allows linkage between various national approaches, covering both direct emissions management and the need to offset emissions.

In such a model, signatories to an agreed global emissions trajectory develop national emissions management programmes. A signatory may then choose multilateral participation in the global carbon market by accepting, at the national or sector level, a fixed carbon emissions budget for a given future period. The budget arises from the goals of the specific policy program(s) as a contribution to the global trajectory.

Alternatively, a signatory may choose to begin the task of managing emissions without participating multilaterally, but instead engage in international trade through unilateral recognition of project mechanisms. In cases where national or sector programmes are not forthcoming, the project mechanism(s) should be used as a minimum contribution. Participation in this model is not mandatory, but once in, participants must fulfil their commitments and meet their agreed emission budgets. Participation would always be dependent on a review of the carbon budget submission by an oversight body, as might be established by the UNFCCC.

A specific oversight body is charged with overseeing the development and operation of the necessary infrastructure to allow the market to function.

- Establishing the framework within which projects and national programmes can link.
- Creating the global carbon-trading instruments that will underpin the linkage process and issuing those instruments in response to submitted projects and programmes. Today's instruments consist of elements such as Assigned Amount Units (AAU) and CERs.

- Developing any measurement, reporting, and verification rules associated with the issuance of instruments and the later step of annual reconciliation.
- Developing and issuing guidelines for the basic structure of national programmes. National programmes designed along similar lines will facilitate linkage.
- Developing and operating the necessary international registries for linking (for example, an expansion of the current International Transaction Log, or ITL).
- Governing the overall framework, including overall trade monitoring and reconciliation at the end of each trading period.
- Assessing submissions for inclusion in the international framework.
- Expanding and operating the project mechanism(s). The project mechanism(s) must be revised to cover a programmatic approach and new mechanisms introduced to widen the range of accepted mitigation approaches.

A multilateral linking approach, combined with absolute national emissions targets and a robust project mechanism(s) is key to a global carbon market. Unilateral recognition of project mechanisms will also play an important role, while bilateral arrangements may be a less effective pathway to a global market.

Certain specific requirements must be met before different market-based approaches can be linked. Some are structural, in that without them the approaches are simply incompatible or create an allowance flow in one direction only. Others are required to give confidence that the approaches are broadly aligned in their goals.

The sector to which an emissions mitigation programme pertains must be clearly defined, with an accepted measurement and reporting protocol in place that defines the current status or base situation. Ideally the sector should have a clear, documented and accessible emissions history. As the sector grows and changes, the definition must be robust enough to incorporate such change, otherwise leakage will occur as emissions migrate to other, potentially non-covered, sectors. The carbon trading instruments used must have a uniform definition to ensure they will be fungible across jurisdictions and trading schemes.

The sector must have a defined pathway forward, which can be translated into an absolute emissions change. In a rapidly developing economy, emissions may rise in a specific sector, but overall emissions per unit of output should still fall.

Most importantly, the emissions change for the sector must represent a real contribution to the international goal. Two key aspects within such an assessment include the rate of improvement in energy efficiency and the rate of decarbonization. For a developed economy, the improvements required should also result in an overall reduction in emissions. Within a developing economy, emissions may still rise even with significant improvements in these factors.

Business leaders have endorsed market-based approaches to climate mitigation and urge policy-makers to ensure that the new post-2012 framework

- remains market-oriented;
- establishes a global carbon market that links different emissions management programmes and allows for trading between nations, sectors and projects;
- provides predictable incentives and long-term policy certainty;
- expands the carbon market by allowing for new market mechanisms to be developed (that can be linked to a global carbon market); and
- reforms the existing CDM into a family of mechanisms that support various sectoral and policy-based emission reductions in developing countries.

Sectoral approaches

The WBCSD has worked on sector projects for years to analyse sector-specific climate change and sustainable development challenges, find solutions through business actions, and propose policy measures to enhance the business contribution. Business welcomes the innovative thinking about the 'cooperative sectoral approaches and sector-specific action' concept since its introduction in the Bali Action Plan. This has expanded the nationally focused thinking that has dominated climate policy discussion so far, by introducing the notion of internationally coordinated policy that includes the economic and commercial sector. This may help deliver a more manageable approach to addressing the challenge of climate change.

The sectors that make up the global economy (such as oil and gas, and cement and transport) differ widely and a 'one size fits all' policy solution across all industries and countries will not be possible.

The WBCSD is working with member companies to develop a potential framework for an international sectoral architecture under the UNFCCC by answering the following questions.

- What is a sectoral approach seeking to achieve and what is its purpose?
- What issue is it addressing?
- Why is it needed?
- How could it work?

A sectoral approach could represent an important tool for addressing such issues as we move towards a cohesive global policy framework. The key issues include the following.

- Competitiveness and carbon leakage in trade exposed industries as a result of partial implementation of a global policy framework
- Rapidly growing CO_2 emissions from emerging economies (Under the Kyoto Protocol, only certain developed countries have quantified national reduction targets. However, because of the scale of economic development in emerging countries such as China and India, their contributions to anthropogenic CO_2 emissions are rapidly increasing. As such, more defined contributions are required by these economies.)
- Equity in setting national reduction targets. Identification of emissions and reduction potential on a sector-by-sector basis could assist in this process.
- Measurement, reporting, and verification of activities
- Common but differentiated responsibility of Parties, accounting for respective capabilities (For countries that are not able to assume absolute economy-wide targets in the short term and, where feasible and appropriate depending on national circumstances, some specific national sector-based commitments could be considered. Such an approach may offer some developing countries a complementary short-term pathway to emission reductions and help generate momentum towards a more comprehensive framework in the longer term.)
- Managing emissions in sectors where national ownership is not clearly defined; the best known example of this being international shipping

Sectoral approaches and agreements are, therefore, recognized as a useful dimension in the future climate change regime.

From a business perspective, an important use of the sectoral approach is to rebalance the competitive landscape as a result of the partial implementation of reduction efforts within the sector globally. In this regard, the resulting solution must involve a significant majority of global production in both developed and developing countries.

At the same time, a sectoral approach can also be a framework within which business can contribute to emission reductions, through, for example, technology capacity building and the sharing of best practices.

Importantly, sectoral-based action must not in itself seek to achieve an opt-out from mainstream policy development – for example, the development of a global GHG market. Equally, it should not result in a long-term higher cost solution to emissions reduction or a lost opportunity to look along the value chain as a result of focusing only on sectoral 'silos'.

A future international agreement must be sufficiently flexible to encompass such options by allowing certain sectoral approaches to develop within it. Such approaches may deliver the end-game policy structure faster than would be the case without them, in part because large-scale sectoral agreements offering significant emission reductions may result from negotiations between a smaller number of key countries or regions (such as G8+5).

We recommend a global policy framework that

- accommodates a variety of sector activities, agreements and objectives;
- is guided by developed and developing countries that choose to take on a set of activities or enter into a specific agreement;
- is supported by a 'toolbox' of mechanisms that deliver incentives for country and private sector participation;
- follows a clear governance structure;
- meets a set of criteria in terms of environmental effectiveness, compatibility with national and regional policies, and cost effectiveness, among others.

Such a framework may provide an important tool for addressing a variety of issues as we move towards a cohesive global policy framework.

Adaptation

The Intergovernmental Panel on Climate Change (IPCC) has established that climate change will have serious implications for humans and the environment that surrounds us. According to the IPCC Fourth Assessment Report, Climate Change 2007: Synthesis Report, these will include 'altered frequencies and intensities of extreme weather events, [which] together with sea rise, are expected to have mostly adverse effects on natural and human systems.'

WBCSD members recognize this situation and are beginning to plan accordingly to minimize risks, improve adaptive capacity and resilience, leverage new opportunities and collaborate with the global community.

Adaptation is a process through which societies make themselves better able to cope with an uncertain future. Adapting to climate change entails taking the right measures to reduce the negative effects of climate change (or exploit the positive ones) by making the appropriate adjustments and changes.

It is now acknowledged that even if we do succeed in reducing emissions, some climate change impacts are unavoidable because of the level of GHG emissions in the atmosphere. As a result, adaptation will be necessary because temperatures will continue to rise with the attendant short- and long-term impacts that this will bring. There is an urgent need for adaptation assessment in the short term, as witnessed by the increasingly high costs of extreme weather events, compounded by rising population densities, eroding natural protection systems, and aging infrastructure.

Government leadership is emerging and will be essential in establishing clear frameworks for adaptation. For example, under the UNFCCC, governments are committed to cooperating with each other to prepare for adaptation. Several financial mechanisms exist to support this. The UN Adaptation Fund will become operational in 2008 to provide financing for adaptation projects in developing countries that are party to the Kyoto Protocol. Unlike other funds, which rely on voluntary contributions from donor countries, it will be funded through a 2% levy on projects developed under the CDM. A number of bilateral funding agencies in developed countries have also allocated funding to adaptation activities.

Many local, regional and national governments have also undertaken efforts to predict possible climate change impacts in their respective constituencies, and are looking at options for adaptation. Effective business responses can best happen within clear and stable framework conditions.

The implications of climate change will vary across different business sectors. Some sectors and individual companies may face increasing threats, while others may be less affected and some may even benefit. Climate change is an inexact science and the impacts are still not fully understood. For this reason, the effects of climate change are best evaluated on a sectoral and geographic basïs. From a business perspective, climate change is likely to affect the location, design, operation and marketing of infrastructure, products and services. From a human perspective, climate change will have socio-economic implications for workforces and markets.

The business case for adaptation planning is a strong one. Adaptation implementation should be assessed as it could generate tangible and short-term benefits for business operations. It could also yield benefits for local communities. Key drivers of adaptation planning or implementation include: competitive advantage, cost savings, liability management, investor pressure, regulation and community resilience.

Conclusion

Only by working through issues surrounding climate can business consider climate change to be an opportunity rather than a problem. And until a framework is worked out for international cooperation, technology, finance and market mechanisms as well as consensus gained on the role of business in addressing climate change issue, certain questions will always remain. Do we agree on the policy recommendations? What fundamentals do we want to see in market mechanisms in a post-2012 climate agreement? How can we further expand carbon markets to ensure that business is able to take advantage of the best market opportunities to abate emissions? Can we develop new mechanisms to do this? What would these new mechanisms look like and what would they seek to achieve?

The WBCSD will continue to explore ways in which its members can provide input into the international climate discussions to ensure the voice of business is heard and understood, and turning challenges into opportunities.

Time is not on our side, as climate change, poverty, food scarcity, and security concerns threaten the stability of our world. Business has a key role to play in solving these challenges and must continue to be deeply engaged in finding answer. This means new forms of cooperation between governments and business. The good news is that global business is up for the challenge.

6

The last decade—convergence of environmental and financial markets: a practitioner's view

Richard L Sandor*

Market-based mechanisms such as carbon reduction and trading have become widely adopted as greenhouse gas management moves from the confines of corporate environmental compliance departments into the heart of corporate financial planning. This is in response both to strategic business needs and public scrutiny.

Climate change and policy responses to it mean significant and uneven financial impacts on businesses. These include physical, policy, competitive, legal and reputational risks. Environmental issues today are increasingly understood to be part of necessary risk management as investors and analysts are paying closer attention to climate liability, and customer expectations make it a critical part of a good business model.

Applying market innovation is a critical tool for achieving economic and environmental goals. One of the value propositions of carbon markets is the ability to provide cost-effective emissions management tools for businesses, as well as a disclosure mechanism for the market and a transparent path to price discovery. Carbon emission reduction through markets and the price signals sent by the market help change behaviour and help companies to innovate and implement internal efficiencies.

---

* Chairman and Founder of the Chicago Climate Exchange, Executive Chairman of Climate Exchange plc. and a Research Professor at the Kellogg Graduate School of Management at Northwesten University, USA.

It is instructive to look back at the evolution of wealth creation in the 20th century as we look forward to how wealth will be created in the coming decades. We saw major trends such as strong manufacturing in the 1970s, as well as inflationary pressures that led to financial innovation with interest rate hedging products and mortgage instruments. In the 1980s, there was commoditization of interest rates and asset/liability management. In the 1990s, wealth creation derived from commoditization of information, with the tech boom and telecom developments. Looking forward, the 21st century is likely to be driven by the commoditization of the two most important commodities on the planet: air and water. These are our most important resources, and with a planet of 6.5 billion people, they are going to be increasingly scarce. At one time, economists considered these public goods in such large supply that they need not be treated as scare. Today that premise is no longer tenable.

The global market for carbon emissions has the potential to become the largest physical commodity traded market in the world. The World Bank recently reported that the global carbon market tripled in size to $70 billion in 2008 from $10 billion a couple years earlier.

In Europe, the EU-Emissions Trading Scheme (EU-ETS) was started in January of 2005 to implement the Kyoto Protocol. This established the world's largest cap and trade market for emission allowances. That market has evolved and continues to experience tremendous growth.

In the United States, carbon markets continue to gain interest as an efficient tool for reducing and mitigating greenhouse gases. The robust voluntary market created by the Chicago Climate Exchange (CCX) has been joined now by regionally mandated markets. The Regional Greenhouse Gas Initiative (RGGI) is an effort by ten states in the north east that banded together to reduce carbon emissions from electric power companies. December 2009 marked the sixth auction of RGGI allowances, and open interest on the Chicago Climate Futures Exchange RGGI contract – an important measure of the depth of a futures market – is up 434% from the end of 2008.

The CCX started as a demonstration-size pilot programme with 13 members. Today, its membership totals more than 400 companies. Following CCX in 2003, the Chicago Climate Futures Exchange (CCFE) was launched in 2004. The CCFE, regulated by the Commodity Futures Trading Commission, offers standardized, cleared futures and options contracts for more than 20 environmental products.

In a market-tested example of how carbon may trade under a federal mandate in the United States, CCFE lists a contract called the CFI-US, which requires, starting in 2011, delivery of mandatory federal CO2 emission allowances. In the event that legislation has not yet passed in the United States, the contract would require delivery of another mandated allowance, such as a RGGI allowance. Open interest in CCFE's CFI-US futures contract has now surpassed 1.4 million tonnes.

Globally, carbon trading has expanded over the past ten years under the umbrella of the Climate Exchange plc, a publicly traded company listed on the London Stock Market AIM division (CLE.L). The European Climate Exchange launched in 2004, and is now the leading exchange in the EU-ETS. Last year, CCX entered a joint venture with the China National Petroleum Corporation (CNPC) and the City of Tianjin to establish the Tianjin Climate Exchange—the first emission trading system in China. CCX recently partnered with the People's Bank of China and CNPC to establish a new centre – the China-US Low Carbon Finance and Development Research Center – that will conduct research on financing of low-carbon initiatives. This will include large-scale demonstrations of market-based methods for addressing environmental challenges in China, as well as improving energy efficiency.

In India, the India Climate Exchange (ICX) is under development. Like the CCX, the India Climate Exchange will be designed through a consensus process by India-based stakeholders, with the goal of determining the most cost-effective greenhouse gas mitigation measures. Chairman of the Intergovernmental Panel on Climate Change (IPCC) and head of TERI, Dr R K Pachauri, serves as the Advisory Board's Honorary Chairman.

The ability of a variety of actors to participate in carbon markets – from farmers in Iowa to villagers in India – will continue to provide opportunities and benefits globally. With the right institutional framework, rural environmental initiatives can be linked to the global carbon market. For example, by using clean renewable biogas for cooking fuel, which displaces use of fossil fuels, poor rural households in the South Indian state of Kerala are earning additional income and helping create sustainable local environment by trading carbon on the CCX. Each household, which displaces about 3–5 metric tonnes of CO_2 per year, is a great example of rural empowerment and bottom-up market linkage.

Carbon markets will continue to make a positive impact in incentivizing environmentally sound projects in India and around the world. Exchange trading offers the necessary standardization and transparency.

While each country has unique characteristics that come with different demands and needs, what seems to be clear is the importance an environmental market mechanism will play in meeting those demands. Interest in carbon markets is growing globally as they offer a way to achieve better strategic management of energy costs, new sources of revenue, job creation, and poverty alleviation. Going forward, the international dimensions of this will likely to develop on what could be called a 'pluri-lateral' basis. There will be carbon markets in different pockets of the world that are linked through similar contracts, much like you see with crude oil today or like we saw with cotton in the 19th century.

What is also clear is that there is an inexorable trend at work here. The driving force will be the opportunities created. As the contours of global emissions trading continue to emerge, the steps taken in the carbon markets will inform and drive the international momentum that is needed to confront global climate change, a problem that knows no borders.

Endnote

The author would like to thank Brookly McLaughlin for her assistance in the preparation of this article.

Appendix: Global carbon trading*

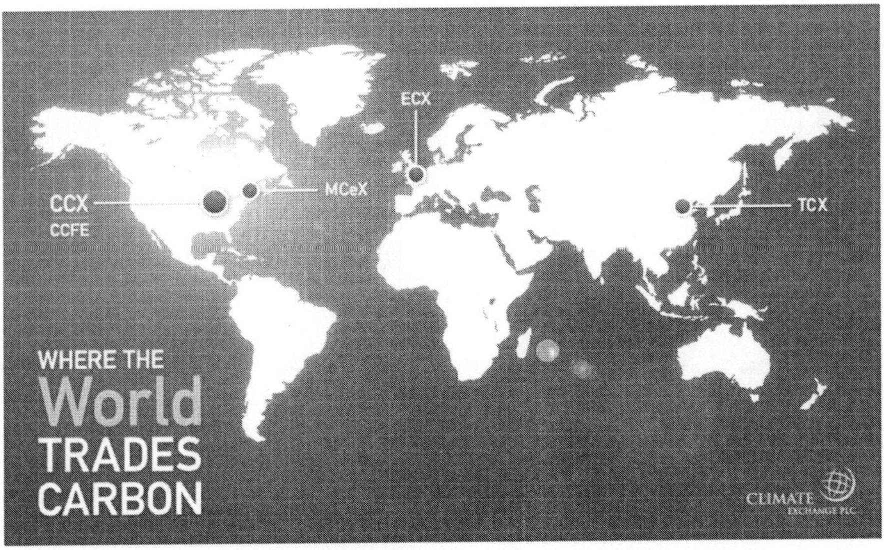

Figure 1 Carbon trading across the world

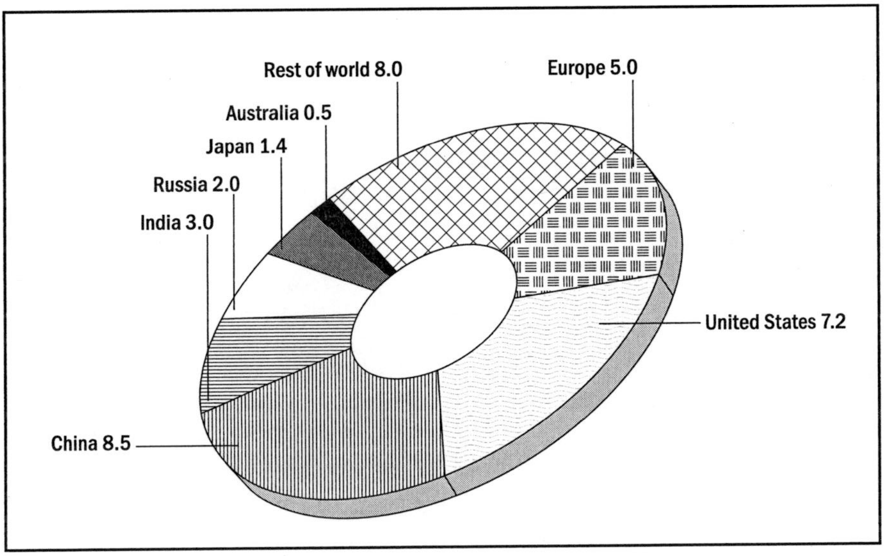

Figure 2 Global carbon emissions (billion metric tonnes)

---

* **Source** Chicago Climate Exchange

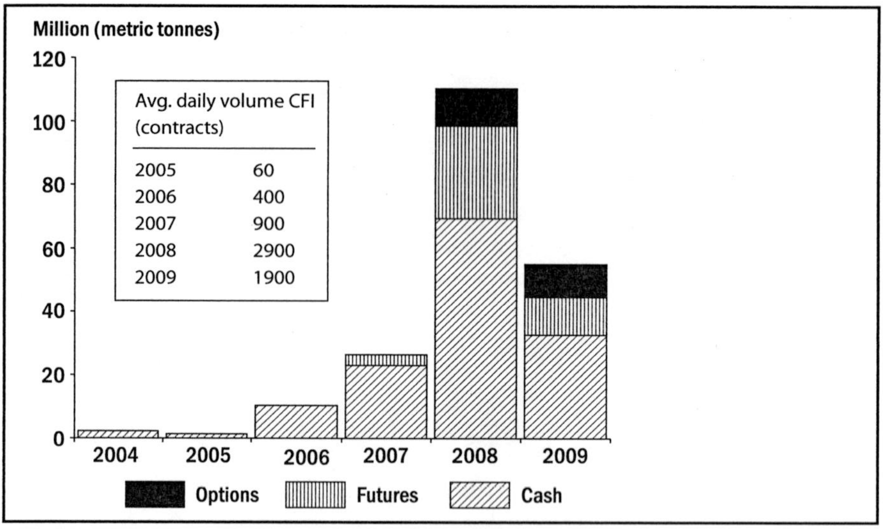

Figure 3 Chicago Climate Exchange (CCX): historic volumes

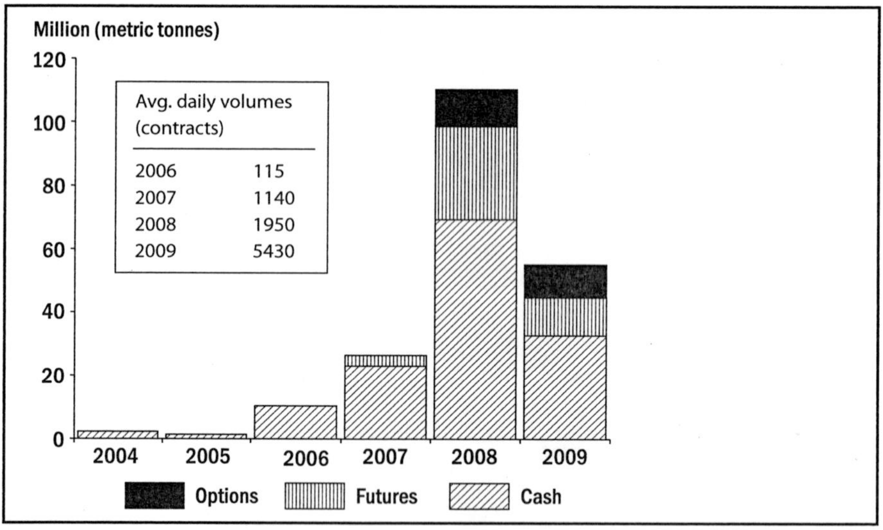

Figure 4 European Climate Exchange (ECX): historic volumes

7

The global climate change regime: a time for consolidation or fundamental reform?

Farhana Yamin and Joanna Depledge*

Introduction

The climate change regime is centred around the 1992 United Nations Framework Convention on Climate Change (UNFCCC) and its 1997 Kyoto Protocol. These two treaties, variously amended, have given rise to hundreds of supplementary decisions that together constitute the international climate change regime—the main set of rules, institutions, and procedures that states have agreed to coordinate their actions at the global level (Yamin and Depledge 2004).

Although climate change has been consistently on the agenda of high level discussions at, for example, the G8 and the UN, and dominates the media headlines, international negotiations to strengthen the UNFCCC and the Kyoto Protocol appear to be moving at a snail's pace. Successive reports of the Intergovernmental Panel on Climate Change (IPCC) have generated a vast body of knowledge, supporting the need for prompt action to curb greenhouse gas (GHG) emissions and enhance resilience and adaptation (IPCC 2007a, b, c). Yet emissions continue to grow in large parts of the world, including in many countries that signed up for Kyoto targets (UNFCCC 2008b). Investment and financial flows of the scale needed to support

* Farhana Yamin, Children's Investment Fund Foundation, former Fellow at Institute of Development Studies, University of Sussex; Dr Joanna Depledge, Associate, Department of Politics and International Studies, University of Cambridge.

adaptation and to move from carbon-intensive business-as-usual (BAU) pathways to climate-resilient development have not been secured (UNFCCC 2007), even though an array of authoritative economic analyses suggests the sums involved are modest when compared with the damages that could result from unchecked climate change (IPCC 2007c; Stern 2007).

What has gone wrong? Why despite nearly two decades of international global cooperation has progress to reduce emissions and help those vulnerable to climate change cope better been so slow? Are negotiators simply out of touch with global climate science? Is their ability to implement solutions thwarted by unavailability of the right conceptual framework(s) and policy solutions that would command widespread public support? Or is the architecture of the UNFCCC and the Kyoto Protocol itself fundamentally flawed, as argued by some (see for example, the views of various authors in Helm and Hepburn 2009)? This chapter attempts to answer these questions by mapping out the achievements and the shortcomings of the international climate regime to date.[1] We then assess in what ways the regime needs to be strengthened in the Copenhagen Climate Summit negotiations, given that the GHG reduction commitments set out in the UNFCCC were judged by all countries back in 1994 to be 'inadequate', and the resulting 1997 Kyoto Protocol only contains GHG-related commitments for industrialized countries up to 2012 and does not apply to the US.

This chapter argues that climate negotiations are at an evolutionary crossroads. To gain credibility, the UNFCCC and Kyoto regime must be strengthened to close not only the gap between climate science and climate policy but also the implementation gap that is growing between agreed policy and its tardy implementation. But doing so will require negotiators to shed at least some of their historical 'baggage' in the form of outdated mindsets and intransigent national positions, as only this will enable the regime to tackle two fundamental, related challenges.

1. How to broaden and deepen participation in commitments, particularly those relating to mitigation.
2. How to better support vulnerable groups and countries with adaptation and climate-resilient development.

---

[1] This article was finalized in October 2009, prior to the climate meetings in Copenhagen. It draws upon and updates views previously expressed by the authors in 'The global climate regime: a defence', in *The Economics and Politics of Climate Change,* edited by Dieter Helm and Cameron Hepburn, 2009, Oxford University Press

Tackling these two challenges in earnest will bring in an array of issues relating to development as well as new actors, constituencies, and mechanism that are currently not considered to be 'core' players in the climate regime. These include finance ministries, local government structures, the private sector, the 'third sector' comprising benevolent organizations, funds, and other mechanisms that are involved in the delivery of aid and social welfare. The integration of these linkages, actors, and mechanisms will require climate negotiators to focus on the third fundamental challenge: reforming the regime's institutional machinery to ensure that it can handle the resulting complexity.

The chapter commences with an overview of the historical and institutional context of the climate regime, including a discussion of its conceptual roots and its relationship to developmental and environmental perspectives. It then explains the key achievements and successes in terms of the core commitments and institutions established by the UNFCCC and the Kyoto Protocol, before setting out in detail how the current negotiations aim to meet the challenge of strengthening these commitments and institutions. Suggestions for institutional reform and its follow up are put forward in the final section of the chapter.

The historical and institutional context

The global climate change regime is founded on the 1992 UNFCCC, which is supplemented by its 1997 Kyoto Protocol. The regime and the challenges it now faces cannot be properly understood in isolation from its historical and institutional antecedents. Regimes, even those on entirely new issues, are rarely constructed on a blank slate, and the climate change regime was no exception. It has been shaped, and sometimes pulled in opposing directions, by two distinct framings and institutional reference points.

The first is the framing of climate change as an environmental/pollution control issue, with a focus on a science-driven, targeted approach (on such approaches and relevance to climate change, see Yamin, Smith, and Burton 2006). The environmental framing focuses on the activities/substances that are causing damage to the environment and tries to limit these to 'safe' levels. Environmental approaches rarely pay much attention to adaptation to the problem, as the focus is essentially on halting the damage being done. A variety of policy approaches are used to regulate environmental pollution, notably setting of target- or technology-based approaches, until a 'safe' level is reached. Institutionally, environmental regimes tend to be located in environment ministries that are usually seen as 'opponents' of BAU, over which they have little control. Typically, environmental approaches result in the setting up of expert groups to provide objective advice to policy-makers on what can be considered 'dangerous/safe' levels of pollution and also on the technological/economic feasibility of various policy options. The best international example of a primarily environmental approach is the regime to combat stratospheric ozone depletion.

The second framing sees climate change as an issue integrally linked to the achievement of sustainable development, which throws up a much broader set of conceptual and institutional linkages, domestically and internationally. Climate change emerged as an international issue in the late 1980s and early 1990s, when sustainable development was becoming the defining intellectual framing for deciding how environment and development could be achieved simultaneously, within the context of international cooperation. The term was used to revitalize not only the mandate of the UN but also global cooperation itself – North–South as well as East–West – at a time when older ideological formations were crumbling (McCormick 1989; Birnie 1993). A sustainable development framing, in particular, highlights issues of intra- and intergenerational equity and North–South dimensions, emphasizing that developing countries are historically smaller contributors to current global environmental problems, have lower capacities and still have high levels of poverty, the eradication of which remains a priority for developing countries.

Because of the wider appeal of the sustainable development framing, from its very inception, the climate change regime was firmly located in the existing UN system. Negotiations on a framework convention were launched by the highest political organ of the UN, the General Assembly (UNGASS), in 1990, prompted by the scientific findings of the IPCC, which was established two years before by the World Meteorological Organization (WMO) and the UN Environment Programme (UNEP). Although there was pressure to set up the negotiations under the UNEP (as had been the case with the ozone regime and the Convention on Biological Diversity), the wider linkages, development dimensions, and higher political stakes of climate change ensured that negotiations took place under the aegis of the UNGASS (Bodansky 1993).

Although climate change now appears to command a disproportionate share of media and UN attention, it is important to recognize that climate change was one of the five major outcomes negotiated as part of the landmark 1992 UN Conference on Environment and Development (UNCED), which was formulated following the recommendations of the 1987 Brundtland Report of the World Commission on Environment and Development. The UNCED resulted in the adoption by all members of the UN of the following documents, which were designed to 'fit' together to secure sustainable development: the Biodiversity Convention, negotiated under the auspices of the UNEP; Agenda 21, a 40-chapter blueprint to put sustainable development into action at multiple levels, with the support of major groups; the Rio Declaration on Environment and Development, a charter defining the rights and responsibilities of states on environment and development; and the Global Principles on Forests (Sands 2003). A newly created body, the Commission for Sustainable Development (CSD), was set up to oversee UN efforts to move towards sustainable development and the newly established Global Environment Facility (GEF) was restructured to ensure that it could serve multiple conventions and processes and help provide holistic, integrated solutions rather than the inefficient, piecemeal approach that would have prevailed if each new instrument had had its own smaller, discrete fund. It is also important to remember that the UNCED gave rise to an array of additional legal instruments, which were adopted after Rio to close the 'gaps' for implementation in special subjects/regions that had not been possible to deal with in the run-up to UNCED, notably in the context of climate change, the 1994 UN Convention to Combat Desertification in countries experiencing drought/desertification, especially in Africa. A high-level follow-up process to monitor the implementation of the UNCED results was also agreed in Rio, involving a five-year review in 1997 in New York and 10-year review in Johannesburg in 2002.

This broader UN setting and the simultaneous UNCED negotiations have ensured that climate change was considered from the outset in a broader developmental context, rather than, as had been the case with the ozone issue, in a predominantly environmental one. As a result, the UNFCCC has a structure that is deeply influenced by a broad range of sustainable development 'principles' such as the principle of 'common but differentiated responsibilities and respective capabilities' (CBDRR). The Convention not only has numerous references to equity but also puts the burden of acting first on developed countries on account of their greater resources and greater contribution to the climate problems.[2] Its provisions are structured to allow developing countries to prioritize poverty eradication and social development, and the countries are entitled to funding and technology from the Convention's own financial mechanism to achieve climate goals in the context of their national development goals.[3]

The formative years of the climate change regime were thus heavily influenced by two quite different framings, with their distinctive mindsets and solution pathways. Negotiators would always look to the recently concluded negotiations on the 1987 Montreal Protocol on Substances that Deplete the Ozone Layer, as this was one of the first treaties that addressed a global problem with significant North–South equity dimensions on the basis of the precautionary principle (Rowlands 1995). But there was always recognition that with no quick technological fix in sight (as there was with ozone), climate change would prove to be a bigger, more complicated problem that implicated development itself—in the North as well as in the South.

Political dynamics

If the North–South divide was the only significant political dynamic in climate change, negotiations would have been a lot simpler than they are at present. As recognized early on, climate change is too big and complicated an issue to permit neat classification. Its causes and consequences impact all sectors, notably, energy, agriculture, water, forestry, health, and tourism, to name just a few. Unsurprisingly, states have divergent interests, depending on the scale of damage and the degree of culpability, and their interests cut across traditional political groupings. As a result, climate change has thus had to deal not only with North–South dimensions, but also the complicated and fractured politics of divisions within the North as well as South

---

[2] UNFCCC, Preamble and Article 4.2

[3] UNFCCC, Articles 3, 4.1, 4.3, 4.4 and 4.5.

(Bodansky 1995, p. 431). Some of the greatest friction has, in fact, come from the disagreements within the Organisation for Economic Co-operation and Development (OECD) countries about, for example, targets and timetables and the use of command and control/taxes as policy instruments versus the merits of cap and trade (O'Riordan and Jaeger 1996; Gupta and Grubb 1999). And, of course, since 2001–08, it was the divisions of approach between the US, arguing essentially for technology and adaptation framing not necessarily focusing on targets versus the targets/environmental framing favoured by the EU and developing countries, that most severely tested the regime. The rejection of Kyoto by the US brought Russia and the political dynamics of East–West relationships briefly to the centrestage, though the enlargement of the EU and Russia's ambivalence to climate politics has made them and the Eastern European block less visible.

The politics of the South, meanwhile, is historically more consistent but no less complicated. Developing countries have always negotiated as a loose bloc within the G77, which China associates itself with. The position of the G77 has been driven by the four main factions. The Least Developed Countries (LDCs) are predominantly in Africa, for which climate is an additional source of vulnerability. Some of the LDCs are also members of the Alliance of Small Island States (AOSIS), which represents 43 small developing country island states for which climate change, in particular, sea level rise, is a matter of survival. The third bloc comprises the relatively wealthy and higher carbon emitting oil-producing/exporting countries (OPEC), which have until recently rejected climate science and favoured inaction. The final sub-grouping comprises essentially political, rather than institutional ties, and comprises the newly industrializing countries (NICs), in particular, Brazil, India, China, and South Africa, the emissions of which, although relatively low on a per capita basis, are rising rapidly. Regional groups, such as the African group and Group of Latin America and Caribbean Countries (GRULAC), can and often do take up leadership positions from time to time, but serve mainly to nominate people for elected posts. The decision by the African Union to appoint Ethiopian Prime Minister Meles Zenawi to represent Africa in the Copenhagen negotiations may be an early indication that regional groups are starting to rethink how to improve representation, especially at the high level political summits that are now playing such a key role in shaping the climate regime .

The relative strength of these groupings within the G77 can change, depending on their political priorities, the personalities, skills and resources of their negotiators and whether stronger allies can be found within or outside the G77 bloc on particular issues. The shared history of colonization and exclusion from the institutions that have more exclusive membership, such as the G8 and the OECD, give the G77 a shared mindset and a need to work together simply to ensure efficacy of negotiations (many countries of the Group have very small delegations that can only manage to make sense of complex negotiations because of the G77 meetings). Despite its centrifugal tendencies and occasional public disagreements over the last 20 years, the G77 has held together and espoused common positions favouring a legal, multilateral framework for climate change, arguing strongly that developed countries take the lead on GHG reductions as well as on the provision of finance and technology while respecting the principle of sovereignty over natural resources.

It is important to bear in mind that the environment/development framings have had supporters that cut across geo-political boundaries, and this has influenced the political and negotiating dynamics on climate negotiations. At first sight, the environmental framing best crystalizes the approach taken by the North, particularly European countries. Environmental NGOs and AOSIS have also consistently called for targets and timetables from the outset. The developmental framing has drawn support from developing countries, the US, international financial institutions, and the donor agency delegates among the developed world. These groups find common ground in stressing the efficiency of a 'development first' approach over a purely environmental one. But there are many examples of countries choosing different framings at different moments in time. For example, the first Conference of the Parties (COP-1) in Berlin saw the emergence of the 'green group', which included the EU, AOSIS, and nearly all developing countries, arguing for what came to be known as 'Kyoto' style targets and timetables. In recent years, the developmental framing has come to the fore once again, with some stressing the links to climate justice and equity issues[4] and others taking a more technical linkage with the broader development agenda (Metz and Kok 2008). South Africa's submissions to recent negotiating sessions, for example, aim to provide

[4] See, for example, Development 2008.

a quantified and measurable approach towards using sustainable development policies to support climate actions (Winkler, Hohne, and den Elzen 2008; Metz and Kok 2008). President Obama's approach to action on climate change as part of the broader challenge of providing price stability, energy security, and making available green jobs is reflected in the official title of the draft domestic US legislation.[5] This broad approach, whilst wholly necessary, is likely to be a core challenge for coming negotiations, as we discuss in later sections.

Core commitments, institutions, and procedures

Understanding what has been achieved by the climate regime is critical to understanding how it can be strengthened and/or replaced by new elements at play in the negotiations. Thus, this section sets out the core commitments and institutions of the climate regime, with a view to assessing their strengths and weaknesses while considering the future challenges.

One of the key tasks for regimes is to define the participants and assign rights and responsibilities to them. Both the Convention and the Protocol are global instruments, in the sense that they allow for, and seek, global participation in their governance processes. The model of global participation was derived from the UN setting but has always had critics. A number of early scholars questioned whether a global instrument was necessary or even possible, given the divergent nature of state interests.[6] Stone, for example, asked whether there was one global climate problem or many, given that states would be impacted very differently (Stone 1990). However, the debate was settled relatively early on in favour of a global regime as policy-makers realized that a restrictive regime might provide a competitive advantage to non-participants and would lack international legitimacy (Bodansky 1995). As Hoffmann (2007) persuasively demonstrates, the ozone negotiations also established a norm of universal participation so that by the time climate change found a place on the international agenda, this norm would have already 'pervaded the entire political context'

[5] The official title of the Waxman-Markey Bill is now HR2454, American Clean Energy and Security Act of 2009. The purpose of this bill is described as 'to create clean energy jobs, achieve energy independence, reduce global warming pollution and transition to a clean energy economy.'

[6] See, for example, US Congress, Office of Technology Assessment, 1994.

(Hoffmann 2007, p.130). Developing countries had become newly involved in international environmental politics through the ozone negotiations and wanted to sustain that involvement (Rowlands 1995). Moreover, some were resentful at their lack of real engagement (that they were 'caught napping') in the early ozone talks, and felt that their interests had not been sufficiently addressed as a consequence. They were determined not to let that happen again in the climate change context. In recent years, some initiatives have sought to, in effect, reopen the debate on the global nature of the regime by arguing that it is really only the top 20–30 countries that need to be at the negotiating table, as their collective emissions drive climate politics (Rowlands 1995). This has resulted in a number of bilateral or regional initiatives, with a limited membership, such as the Asia-Pacific Partnership and the Major Economies Initiative set up by former US President Bush, the intent of which appeared to be to replace the global framework created by the UNFCCC and Kyoto. The Obama administration's re-titled Major Economies Forum (MEF) is intended to supplement, rather than replace, the global framework and has created an important new space for more informal and frank dialogue to take place. The MEF has also evolved beyond the seventeen large emitters to include representation from smaller, vulnerable countries, who participated first as observers, and at the October 2009 London meeting as full participants.

Within the model of global participation, the actual commitments of UNFCCC parties are heavily differentiated among developed and developing countries, in keeping with the sustainable development framing discussed above, which gives weight to the aspirations of developing countries. The Convention thus enshrines the principle of CBDRR, establishing that although climate change is of common concern, historical responsibility for causing the problem lies with the developed countries, which also have greater resources to address the problem. The developed countries should thus 'take the lead in combating climate change and the adverse effects thereof'.[7]

This leadership paradigm is translated into the operational rules of the Convention, which require 'developed countries' to take 'the lead' in 'modifying long-term trends in emissions', and includes a list of such countries.[8] These 41 'Annex I' countries include the members of

---

[7] UNFCCC Article 3.1.

[8] UNFCCC Article 4.2(a).

the OECD (in 1990), along with countries of the former Soviet Union and Central and Eastern Europe.[9] The Kyoto Protocol is modelled on the same global/differentiated approach: participation is global, but commitments are (heavily) differentiated, with fixed binding emission targets for the Annex I Parties. Neither the Convention nor the Protocol explicitly states that the 150+ non-Annex I developing countries will follow the lead taken by the Annex I parties. The inference, however, is there. The non-Annex I Parties do have general commitments under both instruments in respect of mitigation and adaptation, as well as reporting requirements. A sub-set of the Annex I countries – 'Annex II Parties' comprising only the OECD members – is committed to providing financial and technological assistance to help developing countries meet their obligations under the regime. They are also obligated to provide funding for adaptation costs for countries vulnerable to climate change.[10] The Convention and Kyoto Protocol also create special rules for LDCs, and give recognition to those considered highly vulnerable, such as the small island developing states (SIDs).

Institutions and procedures

The basic institutions and procedures of the climate change regime are fairly unremarkable, being drawn from existing practice and precedent in the UN system. In terms of institutions, the COP serves as the supreme decision-making body for Convention affairs, and the COP serving as the meeting of the Parties to the Kyoto Protocol (CMP) governs Protocol affairs. The COP and CMP are supported by two permanent subsidiary bodies – the Subsidiary Body for Scientific and Technological Advice (SBSTA) and the Subsidiary Body for Implementation (SBI) – with additional bodies convened for particular negotiating rounds, such as the ongoing Copenhagen negotiations, discussed subsequently. All parties to the Convention are represented in these core institutions, but only Protocol Parties may take part in decision-making on Protocol matters.

The regime also encompasses a raft of highly innovative limited-membership institutions, each with their own mandates and structures.

---

[9] Specifically those that were members of the UN during the Convention negotiations.

[10] UNFCCC, Article 4.4.

Chief among these are the bodies that were established to govern the extensive carbon markets launched by the Kyoto Protocol, namely the Executive Board that oversees the day-to-day running of the Clean Development Mechanism (CDM-EB) and the Supervisory Committee that performs a similar function for Joint Implementation (JI). The Kyoto Protocol also led to the creation of the Compliance Committee, which runs the compliance procedures, along with the more recently established Adaptation Board that oversees the Adaptation Fund, which is financed by a ground-breaking levy on 'credits' resulting from transactions under the CDM. There are also other expert groups under the Convention, with advisory and analytical functions, for example, dealing with the special needs of LDCs, promoting technology transfer, and developing country reporting issues. Most of these were established as part of the Marrakesh Accords, a detailed rule book agreed at COP-7 in 2001, to complete the Kyoto Protocol.

The conduct of business is then governed by formal procedural rules, including those set out in the Convention, Protocol, and the 'draft rules of procedure' (UNFCCC 1996). These determine procedures for intervening in debates, interpretation and translation, establishing informal groups, taking decisions, appointing chairpersons, and so on. They are supplemented by unwritten established practices that have developed over a period of time, both in the UN system in general and specifically in the climate change regime. These procedural rules are distinctive in two main respects. The first is the regime's dubious distinction (which it shares with the Convention on Biological Diversity) in having failed to formally adopt its 'draft rules of procedure', due to disagreement over the voting rule. The effect is that most decisions can only be taken by consensus. The second is the seat on the COP Bureau[11] reserved for SIDs, in recognition of their urgent stake in the regime's work. This measure has subsequently been incorporated into virtually all of the limited membership bodies established by the Convention and the Protocol. As discussed further, the limited membership bodies have their own procedural rules, including novel features quite distinct from the more traditional procedures of the broader regime. The CDM-EB and Compliance Committee, for example, allow for decision-making

[11] An 11-person committee that advises the COP.

by voting, where consensus is not possible, which marks a significant point of departure for the regime.

Key achievements

How has the climate change regime performed as a contractual environment? At one level, it has notched up considerable success, notably in generating political momentum, enabling reciprocal deals, facilitating learning, and promoting reporting and verification.

Momentum

The climate change regime has unquestionably generated momentum, engaging parties, even the more reluctant ones, in a continuous move to advance the issue at hand. Both the 2005 launch of negotiations on second period targets under the Kyoto Protocol and parts of the 2007 Bali Road Map (described in Box 1) were on the agenda because of

Box 1 2007 Bali Road Map

Negotiations to strengthen the climate regime are taking place in several institutional settings in the UN as well as in regional and global political fora. Climate change is now routinely discussed, and elements of the big picture are identified in various UN, bilateral, and regional fora (for example, the UNGASS, G8 plus 5, Asia Pacific Partnership, and the Major Economies Forum). The international negotiations under the UNFCCC and Kyoto will, however, continue to assume prime legal and political positions for the delivery of a future agreement. A number of distinctive negotiating groups have been set up, which are to conclude their work in time for adoption by parties in Copenhagen 2009. These bodies include the Ad-Hoc Working Group on Long Term Cooperative Action (AWG-LCA), established under the UNFCCC, which is looking at national implementation/international actions, adaptation, financing, and technology issues for all parties and the Ad-hoc Working Group on Article 3.9 (AWG-KP) under the Kyoto Protocol, which is looking at the commitments of only developed countries beyond 2012 for countries that are parties to the Kyoto Protocol. Together, the work of the AWG-LCA and AWG-KP is referred to as the Bali Road Map. Another process is dealing with the review of the Kyoto Protocol mandated by Article 9. This picks up wider issues such as the application of a 'levy' to the Kyoto mechanisms. The Convention's bodies are also looking at the inclusion of reductions of emissions from degradation and deforestation (REDD).

reviews and deadlines scheduled in the UNFCCC and Kyoto Protocol. Likewise, the negotiations that led to the Kyoto Protocol were launched in 1995 by a formal review mandated in the Convention.

Momentum, however, is more than diplomatic deadlines inscribed in legal text. Climate negotiations create a dense web of meetings, practices, processes, and rules, so that participation in the regime becomes a profession in itself, generating an inherent motivation among negotiators to advance their issues. The personal connection is significant too. Long-standing climate change delegates, which are many, have evolved into a distinct 'process community' (Sjöstedt 2003. p.102), criss-crossed by personal relationships, sharing a common history and, to some extent, working for a common purpose. This creates a form of peer pressure to reach agreements or at least not to block a consensus. Another form of pressure is the visibility of regimes to the outside world. Media coverage during negotiations can play an important contributory role – although rarely a decisive one – in inducing countries to strike a deal.

The momentum of the climate change regime has been remarkably powerful throughout its history. It was momentum in July 2001 that propelled the climate change parties to recover from the double-blow of a negotiating failure at COP 6 in The Hague (November 2000), and the US repudiation of the Kyoto Protocol a few months later, to strike a political deal on the details of the Protocol in Marrakesh. Undoubtedly, the threat that the entire multilateral framework might collapse focused minds to remember the strengths of an international framework, however imperfect. The adoption of the Marrakesh Accords reflected a clear sense of common determination among the remaining climate change parties to save and advance the regime.

It was also momentum that helped seal the final deal on the Bali Action Plan in December 2007. To recall, the US initially rejected the proposed draft, and then appeared to change its mind and join the consensus, following strong and emotional appeals from other parties (ENB 2007). It is important not to exaggerate the influence of regime-induced momentum, including an inherent sense of community and outside media visibility in this regard. To be clear, the US did not change its mind in Bali because of internal pressure within the regime. What the internal momentum did do, however, is to exert strong pressure on the US and others to intensify their efforts to find a solution. Indeed, the pressure felt by the US has been unremitting since 2001. The continuity of the regime and its forward movement have meant that the US has been utterly unable to ignore the issue of climate change. The effect is that the US has continued to be engaged in the regime, not just under the Convention to which it remains a party, but also as an observer to the Kyoto Protocol. With over 120 members, the US delegation in Bali was larger even than in Kyoto. At another level, several analysts have pointed to the blossoming of actions to tackle climate change at many levels within the US, despite the federal government's rejection of the Kyoto Protocol (Koehn 2008; Pew Center 2006, 2008). Would these many actions have emerged in the absence of the climate change regime? Would the Obama administration be able to put forward domestic climate legislation if it were not for the regime?

The internal momentum of the regime has acted almost as a centrifugal source, keeping other reluctant states involved. Saudi Arabia, other oil-exporting allies, the Russian Federation, Canada (under recent administrations), even China and India—none of these states are inherently progressive about tackling climate change, yet all have remained within the regime and engaged in the Bali Road Map negotiations. The case of Australia is particularly instructive. As soon as the new administration of Kevin Rudd assumed office in late 2007, Australia rejoined the Kyoto Protocol, reflecting the 'pull' exerted by the regime. This momentum was critical to launching the Copenhagen negotiations.

Reciprocal deals

The climate change regime has also led to complex reciprocal deals over a period of time. The launch of the Copenhagen negotiations was itself a classic demonstration of collective decision-making enabled by regimes. The adoption of the Bali Action Plan in 2007, which launched the Copenhagen negotiations, by all accounts represented a decisive political breakthrough in establishing that developing countries would now take on obligations beyond those mentioned in the Convention. This was an unmistakable exercise in complex and deferred reciprocity: in 1992, developed countries agreed to lead; in 2007, developing countries agreed to follow. The strengthening of rules within the regime, including entry into force of the Kyoto Protocol and the flourishing of the CDM, no doubt contributed to the sense that leadership, of a sort, was being exercised by the developed countries. Perhaps most importantly, at CMP 1 in Montreal in 2005, developed countries agreed to launch negotiations on second period targets for themselves under the Kyoto Protocol without, at that stage, any undertaking on the part of developing countries. This represented a tremendous leap of faith in deferred reciprocity, which was, in the end, vindicated in Bali. Learning over a period of time, facilitated in part by the IPCC and exchange of information within the regime, also helped persuade developing countries of the need to move on from their Convention obligations. The scientific point – that the atmosphere does not have boundaries and does not react to equity arguments – has also encouraged them to accept the need for further actions.

Learning

Learning has certainly been an important positive dynamic for the climate change regime, feeding through to the Copenhagen negotiating round. The sheer amount of information and analysis passing through the regime in support of the negotiations is massive. There are several levels of such information. The IPCC provides the fundamental scientific basis for the continuation of the regime and (to a lesser extent) the setting of targets. Although it is, strictly speaking, outside the regime, the IPCC maintains close links to the regime's policy organs, including through provision of common methodologies and data sets for states to work on. In this respect, the climate change regime has proved to be well-designed, safeguarding the IPCC's independence, yet ensuring its relevance. The importance and influence of the IPCC can be seen in debates on second period targets under the Kyoto Protocol, where the Panel's figures have been cited as reference points for their determination. At another level, the regime has received many proposals from parties, seeking to share ideas on creative outcomes for the Copenhagen negotiations. Informal workshops have been held to discuss these proposals further in a non-confrontational environment. Channels for input from civil society, notably side events on the fringes of the official negotiations, have enabled the development of ideas in academia or among NGOs that can be moved onto the political agenda.

Perhaps the most concrete proof of learning in the recent history of the regime lies in the market mechanisms of the CDM, JI, and emissions trading. Ten years ago, in Kyoto, the controversy generated by proposals on the CDM and emissions trading nearly brought the negotiations to collapse. JI was so politically sensitive, it could not even be mentioned by name.[12] Today, the CDM is a massive multi-million dollar industry (notwithstanding its shortcomings), emissions trading is at the heart of the EU response to climate change (with the EU having been one of its sharpest critics in the 1990s), and JI has officially re-entered the vocabulary of the climate change regime (UNFCCC 2006).[13] Most remarkably, opposition to the market mechanisms on ideological grounds is now confined to the very fringes of the climate change debate, although there is a growing recognition that carbon markets can and should be improved to safeguard the environmental integrity of the regime. One of the very few early areas of agreement in the ongoing negotiations on second period targets under the Kyoto Protocol was that the market mechanisms should 'continue to be available' to parties beyond 2012, with the focus very much on how improvements can be made based on experience to date (UNFCCC 2008a).[14]

Another example of learning concerns reducing emissions from deforestation and forest degradation, or REDD. The raising of this issue in the climate change agenda, since its proposal by Papua New Guinea and Costa Rica in 2005 (UNFCCC 2005b), represents a massive step forward in terms of addressing deforestation in developing countries. It was facilitated by methodological developments within the IPCC, research within the academic community, and NGO advocacy linked to the regime, eventually reaching the formal negotiations through concerned governments. The REDD debate in the climate change regime has moved further and faster in two years, than in 15 years of discussion in other forest-specific regimes. Interestingly, it is also a case wherein the substantive incentives provided by the climate change regime, notably market-based credits, have added a new, positive motivation to tackling this issue.

Other new ideas have moved from academic/think-tank debate into the political arena of the Copenhagen negotiations, notably SD-PAMs (sustainable development policies and measures), sectoral

---

[12] 'Activities implemented jointly' (under the Convention) and 'Article 6 projects' (under the Kyoto Protocol) were the terms used instead.

[13] Decision 10/CMP.1.

[14] Para.18.

targets, sectoral-CDM, and levies on air transport to fund adaptation or mitigation activities in developing countries. All of these originated within the research community just a few years ago, but are now on the formal negotiating agenda, proving that the regime is able to rapidly absorb fresh thinking and approaches. On these and other proposals, the regime has provided an open and effective forum for the development and spread of ideas.

Another case of learning lies in the new limited membership bodies set up under the Kyoto Protocol. Acknowledging the inefficiencies of the regime's existing institutions, with their open participation and consensus decision-making, parties decided that the day-to-day governance of the CDM, JI, compliance system and adaptation fund should be entrusted to much smaller, more nimble institutions (maximum of 20 participants). The limited membership bodies may resort to majority voting and have pioneered innovative procedures, such as electronic decision-making. This is an example where the parties explicitly recognized the need to move away from existing practice, and try out new institutional models.

Reporting and review

A major achievement of the climate change regime, and one that is often overlooked, has been to establish perhaps the most rigorous and respected reporting and review system for developed countries of all multilateral environmental agreements (MEAs) (for a more detailed review and response to recent critiques, see Swart, Bergamaschi, Pulles, *et al.* 2007). Since 1996, Annex I parties have been required to submit emissions data on an annual basis, according to common rules and guidelines. This data is subject to individual and collective review, before being published on the convention website and in official documentation. The reporting record, while not perfect, has improved considerably since the early days of the regime, especially among the economies in transition (EITs), where capacity for estimating emissions was limited at first. The reporting and review system has strengthened over time and has now expanded to cater for verification of the legally binding emission targets under the Kyoto Protocol, along with the recording of credit exchanges under the market mechanisms. The CDM has similarly pioneered verification techniques, with the introduction of ex-ante and ex-post verification of emission credits generated by CDM projects through independent, accredited organizations. The use by the regime of private sector entities to perform essential verification functions is a highly novel and significant feature. It demonstrates that, notwithstanding its location within the UN system, the regime is able to make use of new actors and the private sector where appropriate.

The collection and review of data in these myriad ways is critical to building confidence in the regime, providing the basis for assessing compliance with legally binding targets and for ensuring the validity of transactions under the Kyoto mechanisms. The availability of sound emissions data and a well-functioning reporting and review system are now taken for granted in the regime. Without it, however, effective action to tackle climate change would simply be impossible. If it were not for the climate change regime, it is difficult to imagine that reporting and validation of emissions data would have progressed so far among such a wide range of countries. The flipside, however, is that these achievements have yet to be matched by developing countries, as discussed in the subsequent section.

Climate regime at the crossroads: challenges for future negotiations

The climate change regime has unquestionably engaged states in a process of collective decision-making. Even its most ardent supporters, however, would admit that progress in ratcheting-up targets has been slow and, in the case of some important countries, implementation has lagged behind commitments. With climate change negotiations entering into their second decade and climate science telling us we must act decisively in the next 10–15 years to avoid dangerous climate change, the question that needs to be asked is how can the current negotiations contribute to tackling climate change. And will this be sufficient to ensure engagement of the key players: the US, the EU, and the remainder of the OECD countries, as well as key developing country groups?

The slowness of negotiations and tardy implementation by a wide spectrum of parties has been due, to be sure, in part, to the fact that the US has dragged its feet on climate mitigation from the very beginning, in contrast to its early leadership on ozone. Its eight-year disengagement since 2001, while not fatal to the Kyoto Protocol, has undeniably had serious political repercussions on the regime, notably discouraging others, including developing countries, from scaling up their obligations. Few regimes have ever truly prospered in the absence of US engagement. It is to the credit of the EU that it has been able to unite with other EITs and OECD countries and forge a united front on climate change, working with developing countries to secure the entry into force of the Kyoto Protocol and implementing the world's biggest carbon market in the shape of the European Emissions Trading Scheme. This has created pressure on the Obama administration to bring fresh energy and urgency to the climate regime in the form of proposed new domestic legislation, as well as a clear commitment to be part of the international regime. Likewise, Japan's newly elected government has deepened its proposed emission reduction targets in response to international pressure in the negotiations.

Given the more favourable alignment of geo-politics among the OECD, what kind of improvements might be made to the climate regime to strengthen the global basis for cooperation? We argue that improvements can only be made if the underlying nature and scale of three key challenges are understood. The first challenge arises from one of the foundational principles of the climate regime: the concept of CBDRR, which seeks to provide differential treatment for developing countries in pursuit of a common goal, the nature and extent of which are unprecedented in international law. The operationalization of CBDRR has been particularly difficult in the climate regime, blocking progress to widen and broaden the regime. The UNFCCC/Kyoto architecture has not been able to utilize the Montreal Protocol approach, which defined timeframes (a 10-year grace period) and specific objective thresholds (per capita emissions of ozone depleting substances) for when a developing country would need to take on commitments comparable to industrialized countries. These omissions have, in part, generated a high degree of legal uncertainty and mistrust that have slowed the progress on post-2012 issues. The lack of certainty was one of the reasons the US gave for rejecting Kyoto. Developing countries must now play a more constructive role in the regime – it is after all their economies and public that stand to lose most. Action, albeit differentiated according to circumstances, is needed from all sides if the regime is to flourish. And it is this mindset that must change if Copenhagen is to prove successful. The current suite of negotiations has tried to build confidence by not insisting on national caps/Kyoto style targets for developing countries. But many proposals for post-2012 still contain various types of 'graduation' clauses that could prove politically and legally difficult to agree to, given the history of this issue.

The second challenge relates to the development framing which is now widely agreed as being central to the architecture debate. As explained in the introductory sections, the climate regime is itself one element of the global sustainable development framework, created by the UNCED. Despite its deep developmental roots, the climate regime has focused mainly on its environmental aspects of controlling GHG emissions, and the Kyoto Protocol is largely conceived of as an environmental rather than a developmental treaty. A number of scholars and intergovernmental processes have tried to shift the emphasis back to the development framings (Stern 2007; IPCC 2007c; UNDP 2007; Yamin 2005; Metz and Kok 2008). The developmental framing places the climate regime's attempt to induce low-carbon growth as part and parcel of broader efforts to reduce vulnerability and enhance adaptive capacity and resilience. These additional, and necessary, linkages to developmental concerns will bring new actors, momentum, and finance to the climate regime. But all this will also add further legal and institutional complexity to the climate regime. This is because these broader objectives touch on the work of national and international structures dealing with energy, transport, forestry, agriculture, health, and disaster risk reduction to name a few of the linkages currently reflected in the post-2012 proposals. Ensuring policy coherence and institutional competencies, complemented by funding, and effective use of resources across such a broad agenda are fresh challenges to the climate regime, and one that it might struggle to solve in the timeframe available before Copenhagen.

The third challenge relates to the institutional dimensions of how the regime can handle complexity. Climate change is already one of the most complex regimes in the world, encompassing two legally binding treaties, hundreds of COP and CMP decisions, and not to forget now the vast range of detailed rules and decisions emanating from the specialized bodies set up under them, such as the CDM Executive Board, the Compliance Committee, and the various new funds. Although they stand outside the regime in legal terms, the international financial institutions are increasingly playing a bigger role in the regime, particularly in relation to its financing. It is not clear at this stage whether the added complexity of the new negotiations, imbued as they are with expectations that climate will not only solve its own CBDRR-related participation problems, but that it will also contribute to broader social, economic, and development goals, is realistic, given the workloads imposed on negotiations and on delegates.

The need for change: dysfunctional North–South politics

With greater 'harmony' in the OECD camp, perhaps the greatest barrier to effective negotiation in the climate change regime is the persistence of dysfunctional North–South politics. This two-way split is utterly at odds with the reality and complexity of country circumstances relative to climate change reflecting, instead, longstanding ideological differences and historical political allegiances. Even this would not be insurmountable, if there was sufficient trust to enable developed and developing countries to communicate positively and constructively. Instead, negotiations in recent years between the groups have tended to be dominated by suspicion, defensiveness, and misunderstanding, which hinder the rational discussion of proposals (Depledge 2006).

It is impossible to entirely dismiss the suggestion that the design of the climate change regime may be at least partly responsible. Certainly, the regime has made little headway in improving trust among the two major groups. Rather than smoothing it over, the rules of the climate change regime may have contributed to entrenching, and even accentuating, the pre-existing North–South divide. For a start, the Convention does not establish a set timeframe for considering stronger developing country obligations or an objective yardstick to assess developed country leadership. Instead, this is left open-ended, with the obvious result that developed countries seek to bring forward the date when developing countries will assume commitments, and the developing countries seek to push it back. This is in contrast with the Montreal Protocol, which set a clear 10-year grace period, the point after which developing countries would assume commitments comparable to those of developed countries. The fact that the climate change regime did not emulate the ozone precedent in this regard was deliberate, and reflects the stronger leaning towards a sustainable development framework, as discussed before. It also reflects in large part the greater wariness of the developing countries following their experience of the ozone negotiations. Whilst the 10-year grace period for all developing countries across the board is not appropriate for climate change, it is high time developing countries accepted the need to consider more differentiated approaches to separate out different groups of countries and to discuss in what circumstances some of these might take on quantified obligations, whether these are in the form of no-lose, national or sectoral targets or through expanded CDM/carbon markets.

This further differentiation is necessary, given that currently, the climate change regime lumps all 150+ countries that are not included in Annex I into a single legal category of 'non-Annex I' (except for the 49 or so LDCs that are granted special consideration), despite their huge variations in circumstances. In terms of emissions alone, 19 non-Annex I Parties have reported emissions under 1 million tonne/CO_2e, compared with 22 parties reporting emissions over 100 million tonnes/CO_2e (UNFCCC 2005a). The non-Annex I group includes two OECD members (Mexico and South Korea), two EU members (Cyprus and Malta), and several Central Asian republics that consider themselves to be EITs, rather than developing countries. A dozen or so others, particularly the oil wealthy states, have ceased or will in the next 10 years or so cease to qualify for Overseas Development Assistance (ODA), as they will become high-income countries. Interestingly, rules applying to the much smaller group of 41 Annex I Parties are far more differentiated. The EITs are granted a variety of concessions (for example, choice of baseline, exemption from financial commitments), while emission targets under the Kyoto Protocol, as well as subsequent forestry allowances, were all decided on an individual basis for each country. A whole new category was even introduced to accommodate the special circumstances of Turkey. Differentiation has been considered an offensive word in the G77, but there is a need to change the mindset to foster a constructive stance. All countries, not just Annex I, need tailor-made solutions for their specific circumstances. Future differentiation of obligations among non-Annex I Parties is virtually a prerequisite to any sensible engagement of so many vastly differing parties.

A related problem is the lack of dynamism built into the Annex I/non-Annex I structure. Non-Annex I Parties may voluntarily join Annex I, or otherwise assume the commitments of Annex I Parties. A small handful of parties have done so, most of which are politically close to the EU.[15] However, there is no automatic 'graduation' process or objective criteria by which to judge whether countries might be ready to join Annex I. The Montreal Protocol, by contrast, set an automatic threshold of consuming 0.3 kg of ozone-depleting substance per capita, above which developing countries had to forego exemption under the grace period. The problem is even more severe with respect to joining Annex B to the Kyoto Protocol, which lists Annex I Parties and their emission targets. Here, the Protocol's rules were specifically designed to prevent easy amendment of the list, and a proposed process for assuming voluntary commitments was also struck down in Kyoto. Because of onerous entry into force requirements,[16] it is, in practice, virtually impossible for any country to join Annex B under the current rules. The only country that has tried to do so, Belarus, is now stuck in a procedural no-man's land: as an Annex I Party, without an emissions target in Annex B, it cannot meaningfully participate in the Kyoto Protocol.[17]

Differentiation among non-Annex I, however, is not easy. The stark division of countries into Annex I and non-Annex I reinforces and perpetuates the North–South political and ideological divide. It dampens initiatives by individual G77 members, and hampers the emergence of more creative alliances between developed and developing countries that might share similar interests. During the Kyoto Protocol negotiations, for example, small island states, along with some Latin American countries, supported proposals that would have allowed individual developing countries to take on voluntary targets if and when they chose to do so. This was criticized by other major developing countries, and therefore quickly rejected. More recently, South Africa even faced criticism for deciding to limit its GHG emissions on the grounds that this was a betrayal of its African and G77 allegiances.[18] It is remarkable how, in the climate change regime, countries have typically chosen to maintain their traditional political

---

[15] Slovenia, Croatia, Monaco and Liechtenstein joined Annex I in Kyoto in 1997. Kazakhstan declared that it would take on the commitments of Annex I Parties in 2000.

[16] Kyoto Protocol Articles 20 and 21. Three-quarters of the Protocol's Parties must individually ratify any amendment to Annex B before it can enter into force. At present, this amounts to around 130 Parties.

[17] For more on this topic, see Depledge (2009).

[18] South Africa criticized for unilateral commitment to set climate targets, 27 August 2008. Details available at <http://www.afriquenligne.fr/s%10africa-criticised-for-unilateral-commitment-to-set-climate-targets-2008082711580.html>.

alliances, even when their circumstances in terms of climate change are quite different. This can result in the rather bizarre situation of some countries having to support group positions for the sake of political unity, when privately they confess they have more in common with the climate positions of 'opposing' blocs. There are exceptions, with some novel alliances being forged in the climate change regime. The Environmental Integrity Group, for example, includes countries from both Annex I (Switzerland) and non-Annex I (Mexico, South Korea), while the Russian Federation's participation in the OECD-dominated umbrella group also constitutes a non-traditional alliance.

Perhaps the most sobering aspect of the inflexible annex system is that it appears to be actively discouraging non-Annex I Parties from joining Annex I and Annex B. There are a handful of non-Annex I Parties (especially those mentioned above that are not G77 members), who might be tempted into joining Annex I and assuming an emissions target under Annex B. Kazakhstan, for example, has expressed an interest in taking on emission targets under the Kyoto Protocol. The difficulties in doing so, however, would deter any party from going down that road. It is regrettable in the extreme that the rules of the regime are actually serving as a barrier to stronger action on the international stage.

Another damaging repercussion has been the impact on reporting of emissions data by developing countries. Unlike the Annex I Parties, who are engaged in rigorous and regular reporting, the vast majority of developing countries have only reported emissions data once, since the adoption of the Convention more than 16 years ago. The result is that there is no comparable time-series set of accepted emissions data for some 150 parties to the climate change regime. The more limited resources and capacities of developing countries, along with the leadership narrative, meant that the Convention rules for their reporting were more lenient. However, there has been almost no progress since then, for two main reasons. Firstly, the issue of developing country reporting has become bound up with the politics surrounding developing country commitments: developing countries resisted scaling up their reporting, because this was seen as a new commitment. Second, inadequate funding has prevented most developing countries from setting up permanent emissions data collection systems in their own countries. The regime has held various workshops and capacity-building initiatives, but these have been insufficient. This deplorable failure to build up the capacity of most developing countries to estimate and record their emissions is due largely to North–South politics, exacerbated by the regime's divisive narrative.

Can climate change and development be successfully linked?

A basic function of regimes is to help lower the huge transaction costs involved in global negotiations. Problems can arise, however, when the sheer number of participants and issues becomes overwhelming, and complexity begins to impact on the negotiations (Zartman 1994). This is undoubtedly a risk in the climate change context, and notably so in the Copenhagen negotiations.

Each negotiating round in the climate change regime has become more complex than the last. The Kyoto Protocol negotiations were seen, at the time, as immensely complicated. However, they appeared remarkably simple compared with the post-Kyoto negotiations, where detailed rules on the three market-based mechanisms, including compliance, land-use, and forestry, and reporting all had to be devised. Now, the Copenhagen negotiations are charged not only with deciding new emission targets for Annex I Parties, alongside possible revisions to the Protocol's rules, but also a whole new architecture for developing country involvement, and possibly that of the US with the broader developmental/energy-focused agendas. To be acceptable to all interest groups, the Copenhagen package will have to include financial provisions, technology transfer incentives, help for adaptation, new rules on REDD, new reporting guidelines for developing countries, provisions on carbon capture and storage, concessions to oil producers for the potential impacts of mitigation policies, possible sectoral elements, a new 'shared vision' – to name just the central elements on the formal agenda.

More fundamentally, beneath these formal agenda items, the Copenhagen negotiations now have to address the broader development linkages, as developing country issues and obligations move centre stage. In one sense, this will undoubtedly strengthen the regime. Stronger linkages with development concerns bring new actors and potentially more resources and political momentum to the climate table. However, they also risk bogging the regime down in problems that are simply too big, and too deeply entrenched across the entire system of global governance, for climate institutions alone to resolve.

In this respect, there is an urgent need for debate over the appropriate role of the climate regime in the broader aim towards poverty reduction and social and economic development.

Complexity is an intrinsic problem with a global negotiation on such a multifarious and wide-ranging issue as climate change. It is made worse, however, by the absence of majority voting procedures, which might help filter out extreme, frivolous or repetitive proposals. The regime has also been unable to develop an effective process for prioritizing issues, with the result that more and more issues are included in ever more comprehensive negotiating rounds. Complexity arguably torpedoed the last round of comprehensive negotiations in The Hague in 2000. It will, at the very least, prove a major deterrent in Copenhagen.

Unwieldy decision-making

The debating procedures of the climate change regime certainly do not encourage judicious decision-making. Almost all decisions, including quite technical ones, are debated in bodies – either full plenaries or informal groups – that are open to representatives from all the parties. To recall, the Subsidiary Body for Scientific and Technological Advice (SBSTA), the body responsible for scientific, technical, and technological issues, is fully open and imposes no limits or criteria on membership or participation. In practice, it is rare that all the 190+ parties and thousands of delegates are intimately involved in a negotiation on a particular topic, with active participants usually confined to 20–30 core individuals, often less. However, combined with the need to report back to coalitions (for example, the EU, G77), this is enough to make for a lengthy and complicated process. Small drafting groups and private informal consultations are used, but these do not have delegated decision-making power. The situation is not helped by the heavy politicization of the climate change negotiations, which means that even the most minor, inconsequential or highly technical decisions tend to be treated as political ones, requiring lengthy negotiation. Although the introduction of smaller, specialized bodies (dealing with CDM, JI, Compliance, Adaptation, and Adaptation Fund) has helped take some of the detailed discussion 'off' the negotiating floor, there is still too much to be discussed in open ended meetings than should be the case.

Furthermore, almost all decisions have to be taken by consensus. This leads to protracted debates, along with lowest common denominator outcomes, given that a handful of parties – usually the least enthusiastic – can always threaten to block a decision (Haas 2002; Underdal 2002). Although rarely used in practice, the existence of majority decision-making in other UN bodies (common in all MEAs except for the suite of treaties agreed at the time of the UNCED) does force parties with more extreme positions to bargain or risk being marginalized. In the climate context, such parties can, and do, hold up proceedings, which results in deferring of decisions on important matters.

Slow learning

Learning within the climate change regime is sometimes slowed down by the absence of an independent technical body that reports directly to the policy machinery so that it could respond quickly to requests for analysis or even provide policy advice when it is needed. The IPCC is a highly important source of scientific assessment, but is itself an intergovernmental body, whose products must be mandated and approved by plenary meetings. The timeframe for the preparation of even the most targeted reports is at least a year. Moreover, the IPCC has no mandate to provide policy advice, and can only review the existing published literature (rather than having experts put forward their own considered opinions). The climate change secretariat can be a useful source of technical information on some issues, but the imperative for it to remain objective limits its impact. The main source of analysis thus comes from the parties themselves (or civil society) and is, unsurprisingly, often widely contrasting, and cannot command widespread legitimacy. This institutional set-up contrasts, for example, with the leaner expert-based assessment panels of the ozone regime, which can prepare reports within a few months, and even make verbal contributions on technical matters during negotiations. These panels are, at times, also mandated to provide specific policy advice. There are several issues in the Copenhagen negotiations on which independent technical advice and analysis would be useful, such as future emission projections in developed countries, current emission estimates, and projections in developing countries, and the feasibility of options for the land-use change and forestry sector. Unless new institutions are created at Copenhagen it is unclear who will provide this in a timely way with the necessary level of expertise and trust.

Suggestions for reform

There is much in the regime that can be usefully consolidated. There is also room for improvement. In practice, any significant change, however ostensibly procedural, involves negotiation within the contractual environment, invoking the politics that themselves pose barriers to decision-making. For this reason, major institutional reforms and innovations have in the past been made as part and parcel of each of the major 'rounds' of negotiations. For example, COP-1 launched the reporting and review process for Annex I Parties and prompted the reform of the GEF to better serve the financial mechanism. Likewise, Kyoto and the Marrakesh Accords created the bodies now servicing the CDM, JI, and the expert groups dealing with LDC and with technology transfer as well as new funding channels and the Compliance Committee. This suggests that institutional reforms could be part of the Copenhagen deal and beyond.

Undoubtedly, the most important reform would be to remedy the rigidity of the structure of commitments, which shape much of climate policy and politics. The annex structure could be changed to make it easier for non-Annex I Parties to join Annex I and assume an emissions target under Annex B. This might go a long way to reduce – or at least stop further entrenching – the divide between developed and developing countries. The factors that are preventing movement into Annex B are currently being reviewed, and options have been put forward to inject greater flexibility. Other forms of emission reduction commitments that are more appropriate for a wider set of countries than top-down economy-wide targets, such as SD-PAMs, no-lose targets, and sectoral approaches are also now being actively pursued. In one sense, therefore, the regime is displaying a capacity to identify problems, learn from experience, and develop new solutions. Political resistance to changing the rules, however, is strong, notably from powerful developing countries, who prefer not to make it too easy to take on new commitments, fearing they may be 'forced' into policy choices that do not fit their specific circumstances. This illustrates well the interplay between regime design and politics. The annex system has created divisions among developed and developing countries, and that division in turn makes it more difficult to reform the annex system.

In a similar way, negotiators in Copenhagen would need to place emissions data collection in developing countries at the centre of the deal. This will require a substantial (but not enormous) injection of financial, technological, and intellectual resources to ensure that all developing countries are able to set up permanent offices to monitor their emissions. It is worth noting that the emissions profiles of many developing countries are quite different to those of most developed countries, being more heavily skewed towards agricultural and land-use emissions. This will require extra, dedicated support. But innovative sources of financing in the form of auctioning of allowances, taxes on aviation, and extending the CDM levy are available as is the possibility that Copenhagen affords for a better alignment and accountability of donor funds currently operating outside the Convention's financial mechanism and sometimes with little regards to climate change.

Another far-reaching institutional change would be for the Copenhagen deal to mandate the creation of smaller decision-making groups, with clearly circumscribed governance powers. These could address more technical or detailed issues, where multilateral diplomacy is inefficient and unhelpful. Of course, the big political issues would still be addressed in the usual open bodies. A limited-membership board/committee could be envisaged for rules on forestry issues, for example, once a political deal has been struck, and also on reporting and review for (selected) non-Annex I countries, given the need for differentiation. This might help free-up the contractual environment of the COP, CMP, and subsidiary bodies for more efficient political decision-making. The climate change regime has reached a stage when it is simply impossible for all issues to be dealt with effectively through multilateral negotiation in the open bodies. The problem is likely to worsen after Copenhagen, with new rules and processes on additional issues (for example, new funds, REDD, carbon capture and storage) almost certain to be launched. The institutional developments and majority voting procedures implemented by the Kyoto Protocol's limited-membership bodies have been positive for the efficiency of the process, and should be replicated and extended to enable the regime to cope with the complexity of climate change. Much more needs to be delegated to specialized groups, otherwise the regime will drown in its own complexity.

In a similar vein, collective learning could be improved by establishing a new, independent technical body to provide analysis and advice in response to specific party requests, preferably in real time. This body should be expert-based (rather than intergovernmental), with a limited membership kept at a manageable level, and should have a mandate to provide policy advice, on request, on the technical feasibility of various options. It should be flexible enough to provide a response to requests in less than 12 months, as the normal annual cycle of the COP often requires rapid inputs. Obviously, parties will be under no obligation to accept its findings. However, there will have to be sufficient trust and acceptance of the new body for its findings to at least be respected as authoritative. One option could be for it to be an offshoot, or subsidiary, of the IPCC.

It is always tempting to advocate majority voting for a broader set of substantive issues as a means of unblocking political strangleholds, enabling faster and stronger decision-making. The problem is that voting generates its own difficulties. It is inherently alienating, with the defeated party left feeling aggrieved. It can, ironically, lead to deadlocks, when a decisive majority cannot be marshalled either in support, or opposition, to a proposal, and can lead to a party withdrawing altogether or simply choosing to openly not comply. In extreme cases, vote buying and bribery can even emerge. Interestingly, the International Whaling Commission, which has always relied on voting for most decisions, has now amended its procedures to place greater onus on the need to reach consensus. This does not mean that there would be no benefits in introducing more majority voting in the climate change regime. Indeed, voting has already been included in the rules of procedure for the CDM-EB and other boards/committees. If additional specialized governance bodies were established, as suggested above, majority voting should be extended to those bodies. In terms of the COP, CMP, and subsidiary bodies, the majority required for decisions would have to be carefully thought through, and probably weighted in some way, to ensure decisions commanded broad support across negotiating coalitions. Smart voting procedures are now in place in a variety of institutions. The Global Environment Facility (GEF), for example, requires a double majority of both donors and recipients, rather than a simple majority, so that one bloc cannot outvote another (Yamin and Depledge 2004). Voting could also be excluded and consensus explicitly required for certain decisions, for example, those relating to GHG reductions as it would be counterproductive to 'force' countries to take on commitments they could/would not implement domestically. The consideration of such options has been largely stalled in the climate change regime and carries considerable ideological baggage dating back to before the Convention's adoption. Careful consideration would have to be given to thinking about whether negotiations for Copenhagen would be advanced by inclusion of voting issues beyond those necessary for operation of limited membership bodies.

The institutional options above suggest a significant expansion of the regime's institutional architecture. In the authors' view, this is unavoidable and reflects the enormity and complexity of climate change, along with the sheer volume of work and existing institutional density. Eventually, the regime appears inexorably headed towards an expanded, permanent structure, with a number of experts and intergovernmental bodies that meet in almost continuous session, akin to the structure of the World Trade Organization. This might, eventually, encourage governments to establish permanent representations to the climate change regime, resulting in a cadre of interdisciplinary climate experts that would hopefully facilitate more efficient decision-making. Such a vision might appear fanciful now, but the regime has nearly reached a stage of permanent session already: the regime's calendar shows meetings held nearly every week throughout much of 2008/09. At present, such sessions are held all around the world on an ad hoc basis. A standing 'international climate change organization' in a single location would surely be more efficient, less exhausting for delegates and more climate-friendly, in terms of carbon footprint. It would also facilitate high level political input from ministers and Heads of State who will be involved to a much greater extent in the climate regime than has previously been the case.

Conclusions

At this stage in time, the biggest structural challenge to the regime is how to overcome the sharp, structural division between Annex I and non-Annex I Parties enshrined within it, which has, undeniably, slowed down action on climate change. Perhaps 20 years ago, the structuring of the climate change regime in this way was inevitable.

But as developing countries expand and integrate their economies with each other and those in the OECD, and as the scale of their political ambitions as players on the world's stage becomes clearer, it is timely to ask whether the existing factionalism can give way to more constructive modes of collaboration. After all, climate change is a common problem and one that the global community has agreed to tackle together. There are promising signs of change. The Bali Action Plan itself established the principle that developing countries would go beyond their obligations in the Convention, while parallel negotiations are underway to try to remedy the defective process for joining Annex B to the Kyoto Protocol. And both developed and developing countries are coming forward with new forms of proposed commitments and actions.

The key question is whether negotiations under the Convention and Protocol tracks can result in a sensible legal package defining a new regime in the time now available before the Copenhagen deadline. The latest rounds of negotiations in Bonn and Bangkok (August and September 2009), produced few results despite the political momentum generated by the Heads of State Summit convened by the UN Secretary General in New York. The recent MEF meeting in London in October 2009 confirmed that a lot of work remains to be done to build trust between developed and developing countries, with key players doubting whether a legally binding outcome could be agreed in time for adoption at Copenhagen at all.

One major faultline that has emerged is whether the Copenhagen deal should be centred around a patchwork of national pledges, as put forward by the US, or whether the regime should continue with the Kyoto Protocol and its approach of top-down, internationally-recognized targets and timetables that are subject to multilateral oversight and compliance, as favoured by the EU and the G77 and China. The choice masks a deeper, historic divergence on the core issue of what kind of additional obligations developing countries should take on and over what timeframe, as well as the extent to which those obligations should be financed by developed countries and through which financial institutions.

Copenhagen is intended to 'seal the deal' on all the main elements of the Bali Action Plan agreed two years ago. However, as evidenced by recent meetings in 2009, the political conflict among the North, across North and South, and within the South, plagued by complexity and seemingly endless arguments, gives little ground for believing that any detailed legal texts – let alone a set of effective decisions that truly tackle the fundamental regime design issues discussed here – can be adopted in time for Copenhagen. Nonetheless, the Danish hosts are maintaining the expectation that a political agreement will be reached to give guidance to the development and adoption of detailed legal texts over the course of 2010. A crucial factor will be whether the Obama Administration will be able to get its domestic legislation passed in time, which will enable the US to make clear the scale of its emissions reduction and financial assistance package. If it does, it will pave the way for a historic agreement that has seemed out of reach for nearly two decades. But what happens if the US is not ready with its domestic package upon which its re-entry to the regime is based? Would a delay be used by the EU and G77 to unite in defence of the global regime – and even improve it – so as to exert more political pressure on the US? Or would it simply lead to the EU and developing countries blaming each other for the collapse of Copenhagen?

Whatever the current difficulties facing the climate regime, it is clear that those difficulties can only be solved through transparent, multilateral negotiations. Proposals to sideline the existing global regime because it is too ponderous, complex, and political, and focus instead on a core group of high emitters, are simply untenable and miss the crucial point, as pointed out by Ambassador Estrada[19] in a recent article on Copenhagen: big disputes among big countries are not reduced just because you make them meet in a smaller room (Estrada-Oyuela 2009). The global climate change regime is a political and legal necessity, and for these reasons must remain the foundation of the international response to climate change.

---

[19] Chair of the Kyoto Protocol negotiations 1995–97.

References

Barnett M N and Finnemore M. 1999
The politics, power and pathologies of international organizations
International Organization **53**(4): 699–732

Birnie P. 1993
The UN and the environment
In *United Nations, Divided World: the UN's role in international relations*, 2nd edn, pp. 327–383, edited by A Roberts and B Kingsbury
Oxford: Clarendon Press

Bodansky D. 1993
The United Nations Framework Convention on Climate Change: a commentary
The Yale Journal of International Law **18**(2)

Bodansky D. 1995
The emerging climate change regime
Annual Review of Energy and Environment **20**: 425–461

Depledge J. 2006
The opposite of learning: ossification in the climate change regime
Global Environmental Politics **6**(1): 1–22

Depledge J. 2008
Striving for no: Saudi Arabia in the climate change regime
Global Environmental Politics **8**(4): 9–35

Depledge J. 2009
The road less travelled: difficulties in moving between annexes in the climate change regime
Climate Policy **9**(3): 273–287

Development 2008
Climate justice and development
Development **51**(3)

ENB (Earth Negotiation Bulletin). 2007
Summary of the Thirteenth Conference of Parties to the UN Framework Convention on Climate Change and third meeting of Parties to the Kyoto Protocol,
3–15 December 2007

Estrada-Oyuela R. 2009
Copenhagen needs a strong lead negotiator: opinion
Nature **461**: 1056–1057

Gupta J and Grubb M (eds). 1999
Climate Change and European Leadership: a sustainable role for Europe?
Dordrecht: Kluwer Academic Press

Haas P M. 2002
UN conferences and constructivist governance of the environment
Global Governance 8: 73–91

Haas P M. 2008
Climate change governance after Bali
Global Environmental Politics 8(3): 1–7

Haas P M, Keohane R, and Levy M. 1993
Institutions for the Earth: sources of effective international environmental protection
Cambridge, MA: MIT Press

Helm D and Hepburn C (ed). 2009
The Economics and Politics of Climate Change
Oxford: Oxford University Press

Hoffmann M J. 2007
Ozone Depletion and Climate Change: constructing a global response
Albany: State University of New York Press

IPCC (Intergovernmental Panel on Climate Change). 2007a
Climate Change 2007: the physical science basis
Contribution of Working Group I to the Fourth Assessment Report of the IPCC, edited by S Solomon, D Qin, M Manning et al
Cambridge: Cambridge University Press

IPCC (Intergovernmental Panel on Climate Change). 2007b
Climate Change 2007: impacts, adaptation and vulnerability
Contribution of Working Group II to the Fourth Assessment Report of the IPCC, edited by M Parry, O Canziani, J Palutikof *et al.*
Cambridge: Cambridge University Press

IPCC (Intergovernmental Panel on Climate Change). 2007c
Climate Change 2007: mitigation
Contribution of Working Group III to the Fourth Assessment Report of the IPCC, edited by B Metz, O Davidson, P Bosch et al.
Cambridge: Cambridge University Press

Jacob T. 2004
Reflections on the current state of global climate response
Climate Policy 4 (1): 91–97

Keohane R O. 1989
International Institutions and State Power: essays in international relations theory
Boulder, CO: Westview Press

Koehn P H. 2008
Underneath Kyoto: emerging subnational government initiatives and incipient issue—bundling opportunities in China and the United States
Global Environmental Politics 8 (1): 53–77

Mathiason J R. 2004
Keeping the peace: regional or global responsibility?
International Studies Review 6 (3): 453–456

McCormick J. 1989
The Global Environmental Movement
London: Belhaven

Metz B and Kok M (eds). 2008
Climate policy as a way to manage climate change risks
Climate Policy 8 (2) *Special issue*

Mintzer I M and Leonard J A (eds). 1994
Negotiating climate change: the inside story of the Rio Convention
Cambridge: Cambridge University Press

O'Riordan T and Jäger J. 1996
Beyond climate change science and politics
In *Politics of Climate Change: a European perspective*, pp. 346–360, edited by T O'Riordan and J Jäger
London and New York: Routledge

Pew Center. 2006, 2008
Learning from state action on climate change
Climate change **101**: State action
Details available at <www.pewclimate.org>

Prins G and Rayner S. 2007
The wrong trousers: radically rethinking climate policy
[Joint discussion paper of the James Martin Institute for Science and Civilization, University of Oxford, and the MacKinder Centre for the Study of Long-Wave events, London School of Economies and Political Science]

Rowlands I H. 1995
Politics of Global Atmospheric Change
Manchester: Manchester University Press

Sebenius J K. 1991
Designing Negotiations Towards a New Regime: the case of global warming
International Security **15** (4): 110–148

Sands P. 2003
Principles of international environmental law, 2nd edn.
Cambridge: Cambridge University Press.

Sjöstedt G. 2003
Norms and principles as support to postnegotiation and rule implementation
In *Getting it Done: post-agreement negotiation and international regimes*, pp. 89–114, edited by B I Spector and I W Zartman
Washington, DC: US Institute of Peace

Stern N. 2007
The Stern Review. The Economics of Climate Change
Cambridge: Cambridge University Press

Stone C. 1990
The global warming crisis, if there is one, and the law
American Journal of International Law and Policy

Swart R, Bergamaschi P, Pulles T, Raes F. 2007
Are national greenhouse gas emissions reports scientifically valid?
Climate Policy 7 (6): 535–538

Underdal A. 2002
One question, two answers. In Environmental Regime Effectiveness: confronting theory with evidence
Cambridge, MA: MIT Press

UNDP (United Nations Development Programme). 2007
Human Development Report 2007/2008.
In *Fighting climate change: human solidarity in a divided world.*
New York: UNDP

UNFCCC (United Nations Framework Convention on Climate Change). 1996
Draft rules of procedure [FCCC/CP/1996/2]
UNFCCC

UNFCCC (United Nations Framework Convention on Climate Change). 2005a
Sixth compilation and synthesis of initial national communications from Parties not included in Annex I to the Convention
Inventories of anthropogenic emissions by sources and removals by sinks of greenhouse gases. [FCCC/SBI/2005/18/Add.2]
UNFCCC

UNFCCC (United Nations Framework Convention on Climate Change). 2005b
Reducing emissions from deforestation in developing countries
Approaches to stimulate action. [FCCC/CP/2005/MISC.1]
UNFCCC

UNFCCC (United Nations Framework Convention on Climate Change). 2006
Report of the Conference of the Parties serving as the meeting of the Parties to the Kyoto Protocol on its first session, Montreal, 28 November to 10 December 2005. [FCCC/KP/CMP/2005/8/Add.2]
UNFCCC

UNFCCC (United Nations Framework Convention on Climate Change). 2007
Investment and Financial Flows to Address Climate Change
UNFCCC

UNFCCC (United Nations Framework Convention on Climate Change). 2008a
Report of the Ad Hoc Working Group on Further Commitments for Annex I Parties under the Kyoto Protocol on the first part of its fifth session, Bangkok, 31 March to 4 April 2008. [FCCC/KP/AWG/2008/2]

UNFCCC (United Nations Framework Convention on Climate Change). 2008b.
National greenhouse gas inventory data for the period 1990–2006. [FCCC/SBI/2008/12]

US Congress, Office of Technology Assessment. 1994
Climate Treaties and Models: Issues in the International Management of Climate Change
Washington, DC: Government Printing Office

Victor D G. 2006
Toward effective international cooperation on climate change: numbers, interests and institutions
Global Environmental Politics 6 (3): 90–103

Victor D G. 2007
Fragmented carbon markets and reluctant nations: implications for the design of effective architectures
In *Architectures for Agreement: addressing global climate change in the post-Kyoto world*, pp. 133–160, edited by J E Aldy and R N Stavins
Cambridge: Cambridge University Press

Winkler H, Hohne N, and den Elzen M. 2008
Methods for quantifying the benefits of sustainable development policies and measures (SD-PAMs)
Climate Policy 8 (2): 119–134

World Commission on Environment and Development. 1987
Our Common Future
Oxford: Oxford University Press

Yamin F. 2005
The European Union and future climate policy: Is mainstreaming adaptation a distraction or part of the solution?
Climate Policy 5(3): 349–361

Yamin F and Depledge J. 2004
The International Climate Change Regime. A Guide to Rules, Institutions and Procedures
Cambridge: Cambridge University Press

Yamin F, Smith J, and Burton I. 2006
Perspectives on 'Dangerous Anthropogenic Interference'; or How to Operationalize Article 2 of the UN Framework Convention on Climate Change
In *Avoiding Dangerous Climate Change Symposium*, edited by Schellnhuber *et al.*
[Proceedings of the UK G8 Conference on Climate Science]
Cambridge: Cambridge University Press

Zartman W I (ed). 1994
International Multilateral Negotiation: approaches to the management of complexity
San Francisco: Jossey-Bass

8

The climate change debate: view from India

Prodipto Ghosh*

The debate on multilateral action on climate change between the developed and developing countries has been sharply polarized for a long-time. If one may mix a metaphor with a descriptive term, India has been in the eye of this storm since the beginning of multilateral concern on climate change in the 1980s, since it has invariably and forcefully brought in the 'development' and 'poverty eradication' sides of the argument, and seems not to buy into the response from many developed countries that the concern for the preservation of the planet's present climate supersedes the former or that aggressive climate action is consistent with maintaining or even improving growth rates.

The positions of both sides, developed and developing, seem to have become further entrenched over time, in particular since the rejection of the Kyoto Protocol by the US Congress on the stated plea that greenhouse gas (GHG) mitigation action by the US would be negated rapidly, unless key developing countries, that is, China and India (and also Brazil, South Africa, Mexico, and South Korea) also undertook similar actions. The US rejection and the stated grounds for it have subsequently spawned a massive political effort by many developed countries, in particular the EU, Japan, Canada, and more

* Member, Prime Minister's Council on Climate Change, and Distinguished Fellow, The Energy and Resources Institute (TERI), New Delhi; the views expressed in this paper are the author's own.

recently, Australia, to reach a comprehensive global arrangement that would also include the US and these key developing countries. Other developing countries are not as serious a target of these efforts.

The key feature of these key developing countries, distinguishing them from other developing countries, is, of course, their size. Brazil is slightly smaller in size in terms of land area (8.514 million km^2), and its population comprises about 60% (187 million) that of the US (area 9.632 million km^2 and population 299.8 million), the largest developed country. India's population is a little less than four times the US population (1134.4 million), but its land area is only about 34% of the land area of the US (3.287 million km^2; the land under cropping is, however, about the same). China's population is more than four times the population of the US (1313 million), while the land area is about the same (9.598 million km^2).[1] Surely, the argument put forth by developed countries outlines the concern that the rapid growth of the economies of these countries, involving increasing use of energy, would lead to such massive quantities of GHG emissions that no matter how stringent the emissions curbs in developed countries, the planet's climate would, in short order, be at severe risk.

The broad response to this argument, which incidentally is also reiterated by these very countries, is that they are all still very poor,[2] despite the fact that they have experienced high gross domestic product (GDP) growth rates in recent times and that their per capita energy use and consequently, per capita GHG emissions are just a fraction of those of the developed countries. Also, in terms of the accumulations of GHGs in the atmosphere, which is what actually leads to climate change, their responsibility is actually negative (Figure 1). Moreover, endless media coverage has led to several myths in developed countries about the alleged energy profligacy and environmental irresponsibility of these key developing countries.

This chapter presents the gist of the arguments made by India, one of the targeted 'key developing countries', and also broadly reflects the concerns not only of this group, but also of other developing countries, since excepting for scale, India is in most respects pretty typical of

[1] Data sources: Population: UNDP (2008); Area: World Bank (2009a)

[2] Human Development Index (HDI) ranks: Brazil: 70, China: 81, India: 128, South Africa: 121. Source: UNDP (2008)

the poorer half of the developing world. However, we need to clearly understand who exactly is responsible for climate change and what is the quantification of this responsibility.

Responsibility for anthropogenic climate change

It is broadly accepted that the climate change problem commenced with the Industrial Revolution, which was based on fossil energy and which gained momentum around the 1850s. Responsibility for the problem, accordingly, should be assigned from this date.

It is often argued by developed countries that an emissions level of 2 tonnes CO_{2e} per capita globally would be (just) within the world's carrying capacity. This would provide a subsistence level of existence. Any emissions above this level should then be accounted as responsibility for climate change.

Figure 1 presents historical responsibilities from 1850 to the present (extrapolated to 2010 based on current trends) for selected developed and developing countries. The historical responsibility of the US (30.5%) is more than fully compensated by either India (–35.6%) or China (–41.4%). It is a fact that China and India together (–77.0%) are

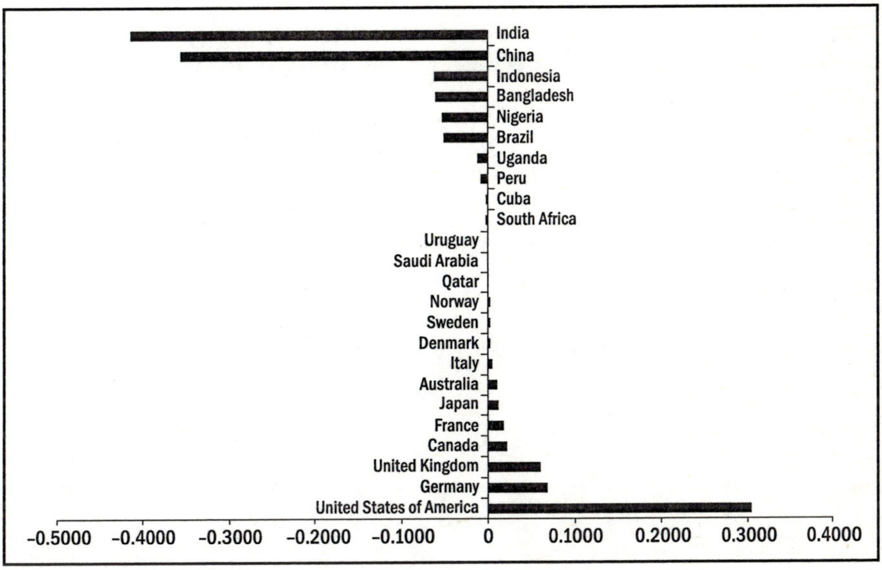

Figure 1 Historical responsibility of selected developed and developing countries (allowing for sustainability level of emissions)

Source TERI analysis (2009)

responsible for the entire environmental space taken by all developed country parties in the aggregate (60.8%) as per the United Nations Framework Convention on Climate Change (UNFCCC).[3]

India's development challenges

Despite several impressive technological progress for example, in space exploration, nuclear energy, information technology, automobile engineering, and agriculture, coupled with a thin margin of prosperity and cosmopolitanism in its teeming metro cities, India remains one of the world's poorest countries. Of the 1 billion+ population,[4] more than 800 million people (79.9% of the population), a population larger than that of North America and EU combined, still subsist on less than $2 per day. Within this population, more than 350 million people, about the population of the US, live on less that $1 per day. More than 700 million people still cook on traditional cook stoves using crop waste and animal residue and a majority of this group, more than 400 million people, do not have access to electricity. One of the enduring images of India's recently concluded general elections is that of the rural poor mobbing parliamentary candidates demanding electricity, so that their children could do their school homework! Electricity enables literacy, as also health care, immunization, safe drinking water, and sanitation, which would enable India to improve its human development indicators[5] (global rank: 128, HDR value: 0.610, life expectancy at birth: 63.2 years, adult literacy rate: 61.0%, GDP per capita: US dollar in purchase power parity: 3452), that are next only to sub-Saharan Africa.

No country in history has improved its level of human development without corresponding increase in per capita use of energy. To expect India to do so would be unrealistic. At present, India's per capita energy consumption is about 20% of the global average, just 4% of the US average, and just 28% of the Chinese average.

---

[3] A 'positive responsibility' would signify that the country has exceeded its sustainability limit of 2 tonnes per capita-year, aggregated over the period 1850–2010. Similarly, 'negative responsibility' would mean that the country has emitted less than its sustainability limit of 2 tonnes per-capita-year over the same period.

[4] Population in 2007: 1123 million

[5] Source: UNDP (2008)

Myths regarding climate change and India

Myth 1: India has done nothing to promote clean energy and energy conservation

Many policy-makers in developed countries acknowledge that while India's need for increased use of energy is legitimate, it should ensure that energy is used efficiently and that clean energy options are employed. With regard to the implications, the Indian government has done precious little on this front.

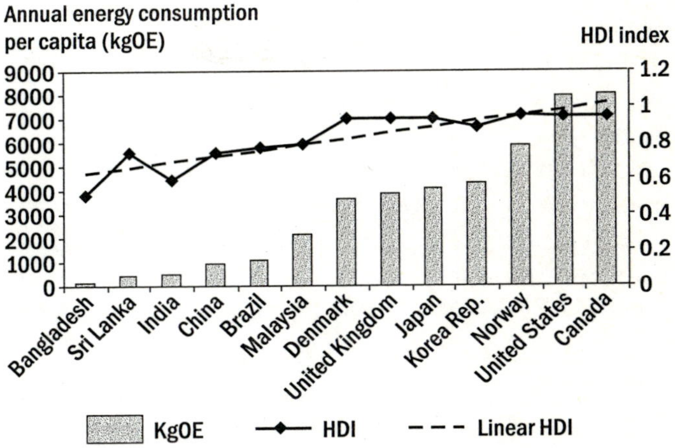

Figure 2 Comparison between Human Development Index and per capita energy consumption of different countries
Note kgOE – kilogram of oil equivalent
Source World Bank (2009b)

As a case in point, India has, over several decades, pursued policies backed by legislation, regulation, tariffs, and publicly funded (almost entirely from domestic fiscal resources) programmes focused on conservation and deployment of renewable energy technologies. Some of these are discussed subsequently.

Reforming energy markets

Reforms have been introduced in the market in the form of legislations like the Electricity Act, 2005; Tariff Policy, 2003; Petroleum and Natural Gas Regulatory Board Act, 2006, and so on. These involve the following.

- Removing entry barriers and raising competition in the exploration, extraction, conversion, and transmission and distribution of primary and secondary energy.
- Instituting price reform, that is, full-scale competition at point of sale; net back pricing for non-traded energy when domestic market is not competitive, and so on.[6]
- Reforming tax regime to promote optimal fuel choices.
- Augmenting and diversifying energy options, sources, and energy infrastructure.
- Providing feed-in tariffs for renewables (solar, wind, biomass cogeneration).

In addition to the above, strengthening or introducing independent regulation has also been done.

a. *New and Renewables Energy Policy, 2005* The policy promotes dependence on sustainable, renewable energy sources, accelerated deployment of renewables through indigenous design, development and manufacture.
b. *Rural Electrification Policy, 2006* The policy promotes renewable energy technologies in places where grid connectivity is not possible or cost-effective
c. *Bio-diesel Purchase Policy* The regulation mandates bio-diesel procurement by petroleum companies.
d. *Ethanol blending of gasoline* The regulation mandates 5% blending of ethanol with gasoline from 1 January 2003 in nine states and four union territories.
e. *Energy Conservation Act, 2001* The legislation aims to reduce specific energy consumption in different sectors and towards this end, a specialized institution BEE (Bureau of Energy Efficiency) has been established for institutionalizing energy efficiency measures, monitoring, and measurement at plant and macro-levels.
f. *Energy Conservation Building Code, 2006* This regulatory code, soon to become mandatory, is designed to ensure energy efficiency

---

[6] In India, energy prices are higher relative to per capita income. The price of 1 million kilowatt-hour of electricity to per capita income is more than Rs 100 compared to 3 cents in the US and 5–10 for most of Western Europe. Similarly, the ratio of the price of 1 million litres of gasoline to gross domestic product per capita is nearly 2000 in India, compared to 20 in the US, and 60–100 in most of the Western Europe.

in all buildings with > 500-kVA-connected load or air-conditioned floor area > 1000 m^2.

g. *Bachat Lamp Yojana (Efficient Lamps Programme)* This is a country-wide programme for the replacement of incandescent lamps by compact fluorescent lamp (CFLs) in households using Clean Development Mechanism (CDM) credits to equate the respective purchase prices. It also ensures safe collection and disposal of used CFLs.

h. *50 000-MW Hydroelectric Initiative, 2003* About 17% of India's electricity is currently generated by hydropower. Additionally, 162 new hydel projects, totalling 50 000 megawatt (MW), have been identified for project preparation, and several are already under implementation.

i. *Several other programmes* These include promotion of solar thermal water heaters, solar photovoltaics (PVs), wind power generation, biomass gasifiers, biogas and manure management, promotion of fuel cells, energy recovery from urban wastes, and so on.

In addition, the Integrated Energy Policy, as an overarching framework, was adopted in 2008. Key provisions of the policy include the following.

- Promotion of energy efficiency in all sectors
- Emphasis on mass transport
- Emphasis on renewables, including biofuels and fuel plantations
- Accelerated development of nuclear and hydropower technology missions for clean energy
- Focused R&D on several climate-change-related technologies.

In addition, in specific economic sectors, a number of policies and programmes have important positive implications for GHG mitigation. Several of these are schematically presented in Table 1.

Myth 2: India remains an energy profligate

Regardless of the actions by India to promote energy efficiency and clean energy choices, an impression persists among many scholars and policy-makers in developed countries that these actions have not 'worked' or are inadequate or are poorly implemented, with the consequence that India is a profligate user of energy, in the sense that

Table 1 Indian policies and programmes impacting greenhouse gas mitigation

A schematic representation of polices and programmes of Government of India that have direct/indirect thrust on GHG mitigation

Sector		
Energy	Policies	▪ Integrated Energy policy ▪ Electricity Rules, 2005 ▪ Tariff Policy ▪ Coal Conservation and Development Act, 1974 ▪ Petroleum and Natural Gas Rules, 1959 ▪ The Oilfields (Regulation and Development) Act, 1948 ▪ Biodiesel Purchase Policy, 2005 ▪ The Electricity act, 2003 ▪ Rural Electrification Policy (approach to rural electrification) ▪ Energy Conservation Act, 2001 ▪ Integrated Coal Policy, 1996 ▪ Petroleum and Natural Gas Regulatory Board Act, 2006 ▪ New and Renewable Energy Policy
	Programmes and schemes	▪ Distribution Reforms and Upgradation Management (DRUM) ▪ Bechat Lamp Yojana ▪ Remote Village Electrification ▪ Solar thermal Energy Programme ▪ Solar Energy ▪ Biomass Power Programme ▪ National Biogas and Manure Management programme (NBMMP) ▪ Hydrogen energy (R&D) ▪ 50000 MW Hydro Electric Initiative ▪ Energy Recovery Power Generation from Industrial and Commercial Wastes and Effluents ▪ National Gas Hydrate Programme (NGHP) ▪ Ethanol blended Petrol Programme ▪ Integrated Rural Energy Programme ▪ Village Energy Security Programme ▪ Solar Photovoltaic (SPV) Programme ▪ National Biogas Programme ▪ Biomass Gasifier Programme ▪ Biomass Energy and Co-generation (non-bagasse) in Industry ▪ Fuel Cells (R&D) ▪ Energy Recovery from Urban Wastes ▪ New Technology Group for the year 2006-2007 (continuation of programmes/schemes-MNRE) ▪ Special Area Demonstration Programme (SADP)
Environment	Policies	▪ Charter on 'corporate responsibility for environmental protection (CREP)' ▪ The Municipal Solid Waste (management and handling) Rules, 2000 ▪ Recycled Plastics Manufacture and Usage Rules, 1999 (as amended 2003) ▪ Batteries (Management and Handling) Rules, 2001 ▪ Forest (Conservation) act, 1980 ▪ Ozone Depleting Substances (Regulation and Control) Amendment rules ▪ Bio-Medical Waste (Management and Handling) Rules, 1998
	Programmes and schemes	▪ Mangrove Conservation Programme ▪ National Afforestation Programme (NAP) ▪ Eco-Development forces (EDF) ▪ Ozone Cell ▪ Regeneration and Eco-development ▪ Grant-in-aid for Greening India ▪ Joint Forest Management cell (JFM)
S&T	Programmes and schemes	▪ *Science and technology application programme* a) Science and Technology Application for Rural Development (STARD) b) Science and Technology Application for the Weaker Sections (STAWS) ▪ *Integrated Village Development* ii, Coordinated Programmes iii, Non-Edible Oils (NEO) ▪ *State Science and Technology Programme (SSTP)* Technology Development Programmes Technology Systems Programmes Joint Technology Projects Under Science and Technology Advisory Committee Mechanism (STAC/IS-STAC) ▪ *Technology Business Incubators*
Housing and urban development	Policies	▪ Constitution (Seventy-Fourth Amendment) Act 1992 ▪ Model Municipal Law
	Programmes and schemes	▪ Jawaharlal Nehru National Urban Renewal Mission, JNNURM ▪ Urban Infrastructure Development Scheme for Small and Medium Towns (UIDSSMT) ▪ Tax Free Municipal Bonds
Transport	Policies	▪ Motor Vehicles Act, 1988 ▪ National Auto Fuel Policy ▪ National UrbanTransport Policy, 2006 ▪ Central Motor Vehicles Rules, 1989 ▪ Integrated Transport Policy, 2001 ▪ Pollution Under Control (PUC) 2004
Biotechnology	Programmes and schemes	▪ DBT Biofuels programme ▪ DBT Programme on Environmental Biotechnology
Industry	Policies	▪ New Industrial Policy ▪ Industries (Development and Regulation) Act ▪ Indian Boiler (Second Amendment) Regulation 2006 ▪ National Jute Policy 2005 ▪ National Steel Policy-2005 ▪ Gas Cylinder Rules ▪ Indian Boilers Act-1923 ▪ National Textile Policy (NT x P), 2000 ▪ Micro, Small and Medium Enterprises Development Act, 2006
	Programmes and schemes	▪ Integrated Development Leather Sector (IDLS) ▪ Technology Mission on Cotton (TMC) ▪ Indian Leather Development Programme (ILDP) ▪ "Integrated Technology Upgradation and Management Programme (UPTECH) (Renamed as "Small industry Cluster Development Programme"), 1998 ▪ Technology Upgradation Fund Scheme (TUFS) ▪ Industrial Infrastructure Updradation Scheme (IIUS), 2003 ▪ Tannery Modernisation Scheme, 2000 ▪ Jute Technology Mission (JTM), 2003-04 to 2008-09
Agriculture	Policies	▪ National Agriculture Policy
	Programmes and schemes	▪ Technology Mission on Oilseeds, Pulses and Maize ▪ Inter-Ministerial Task Force on Integrated Plant Nutrient ▪ Management Using City Compost ▪ Integrated Development of Tree Borne Oilseed by NOVOD Board
Land resources	Programmes and schemes	▪ Drought Prone Areas Programme ▪ Integrated Wasteland Development Programme (IWDP) Started in 1989/90 ▪ Watershed Development Fund ▪ 'Soil Conservation for Enhancing Productivity of Degraded Lands in the Catchment of River Valley Projects and Flood Prone Rivers' ▪ Desert Development Programme (DDP) ▪ National Watershed Development Project for Rainfed Areas (NWDPRA) ▪ Watershed Development Programme in Shifting Cultivation Areas (WDPSCA) ▪ Western Ghats Development Project (WGDP) and Hill Area Development Programme (HADP)
Water supply and sanitation	Policies	▪ National Water Policy (September,1987)
	Programmes and schemes	▪ Accelerated Rural Water Supply Programme (ARWSP) ▪ Central Rural Sanitation Programme (Total Sanitation Campaign)
Rural development	Programmes and schemes	▪ Pradhan Mantri Gram Sadak Yojana ▪ Special Central Assistance for Tribal Welfare ▪ Rural Housing Programme: Indira Awaas Yojana
Mining	Policies	▪ National Mineral Policy, 1993 ▪ Mines and Minerals (Development and Regulation) Act, 1957

Source FICCI (2007)

India uses much more energy per unit of output (with consequent higher GHG emissions) than would be considered significant in developed countries. The following data will dispel this myth.

Consider first integrated steel plants, one of the industrial sectors involving large-scale energy use (Figure 3).

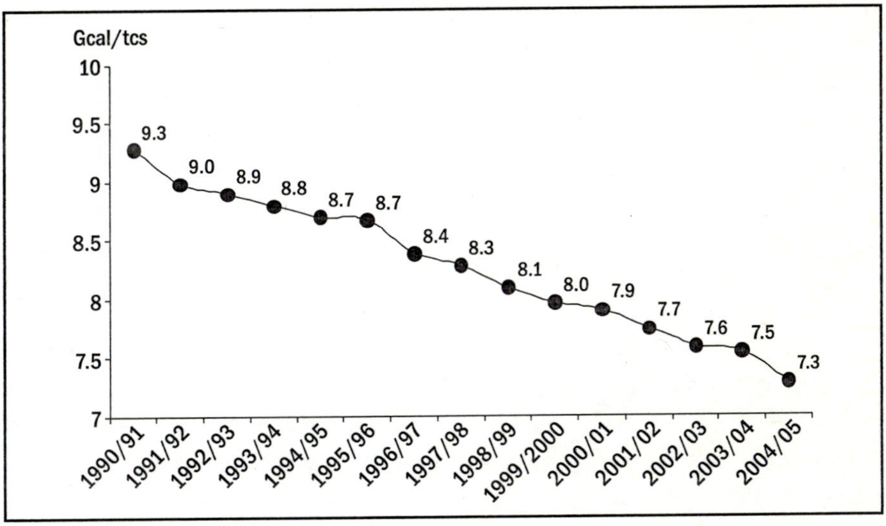

Figure 3 Average specific energy consumption in integrated steel plants
Note Gcal – gigacalories
Source Steel Authority of India (2006)

It would be apparent that in the last decade and half till 2005, specific energy consumption in the steel sector declined sharply. The average decline was, in fact, more than 22%. Accompanying this decline is an increasing share of the 'direct reduction' steel-making technology. It is a fact that the newer Indian steel plants are among the most energy efficient globally.

Consider next, another major energy-using sector, cement. The annual decline in average specific energy consumption in the past decade till 2006 is 7.5% (Figure 4). Two of the most energy-efficient cement plants in the world are located in India.

A similar scenario is presented by the fertilizer sector (Figure 5). The sector has witnessed, on an average, 26% and 25% improvements in specific energy consumption in ammonia and urea plants, respectively, in the last decade and half till 2003. In 2003, the top 25% of Indian fertilizer plants were more energy efficient than the

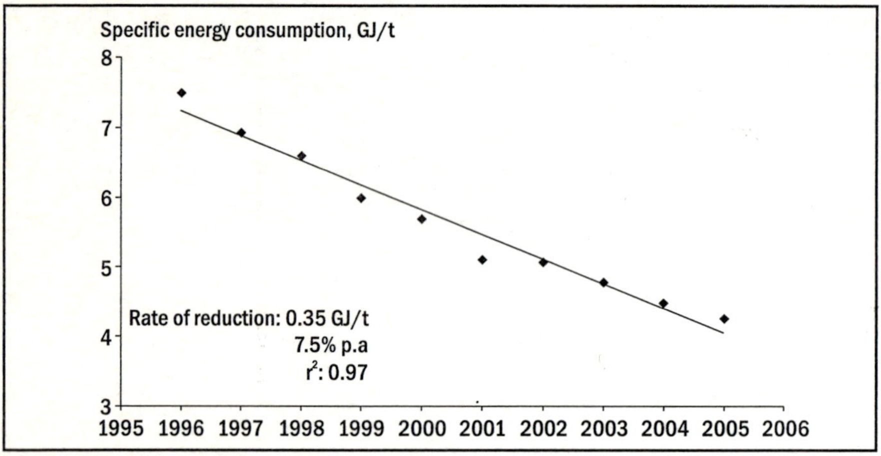

Figure 4 Cement sector energy trends
Note GJ – gigajoules
Source BEE (2007)

top 25% in the world. The best ammonia plant in India had a specific energy consumption of 7.2 Gcal/tonne, compared to the world's best of 7.0 Gcal/tonne. The industry association of the fertilizer sector has set a near-term target of 6.5 Gcal/tonne.

Other major energy using sectors also present more or less a similar picture, the examples being aluminum, paper, power plants,

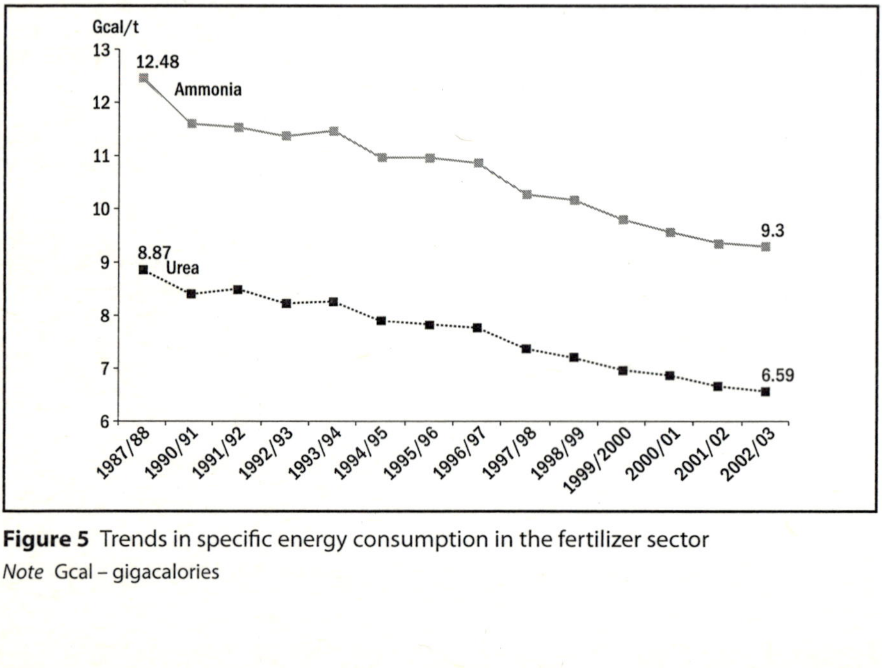

Figure 5 Trends in specific energy consumption in the fertilizer sector
Note Gcal – gigacalories

and petroleum refining (the world's most energy-efficient sector, according to Shell, which is also the world's largest refinery located in Gujarat, India). Since the mid-1980s, India's energy intensity, that is, energy consumed per unit of GDP at purchase power parity (PPP) has declined continuously, and is currently comparable to the EU countries, according to the data from the International Energy Agency (Figure 6).[7]

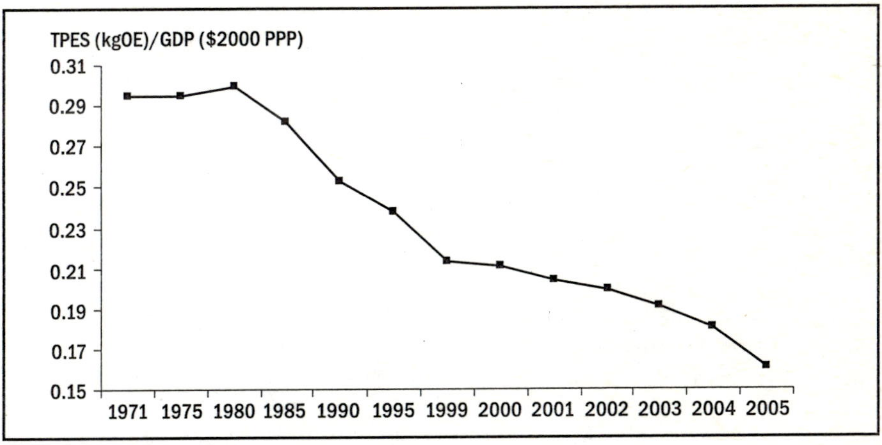

Figure 6 India's energy intensity trends
Note TPES – total primary energy supply; kgOE – kilogram of oil equivalent
Source International Energy Agency, Paris, Database 2006

An interesting comparison was presented by a World Bank study published in 2007, which looked at the carbon dioxide (CO_2) intensities of fossil fuels of the world's 20 largest economies. The findings are presented graphically in Figure 7.

The data is scaled with respect to both CO_2 intensities and GDP per capita in terms of percentage of the corresponding US figure. A brief observation reveals that there is no basis for the common belief in developed countries that they have lower CO_2 intensities per unit of GDP than developing countries. Indeed a formal regression analysis reveals no statistically significant correlation. However, if we note the CO_2 intensity data for India, it is better than that of Germany and about the same as that of Japan, universally cited as the world's most energy-efficient economy!

7 Indian policy-makers assert that the IEA, in fact, overestimates India's energy intensity, by imputing to Indian coal the calorific value typical of European coals, while the Indian coal calorific values are significantly lower due to much higher ash content.

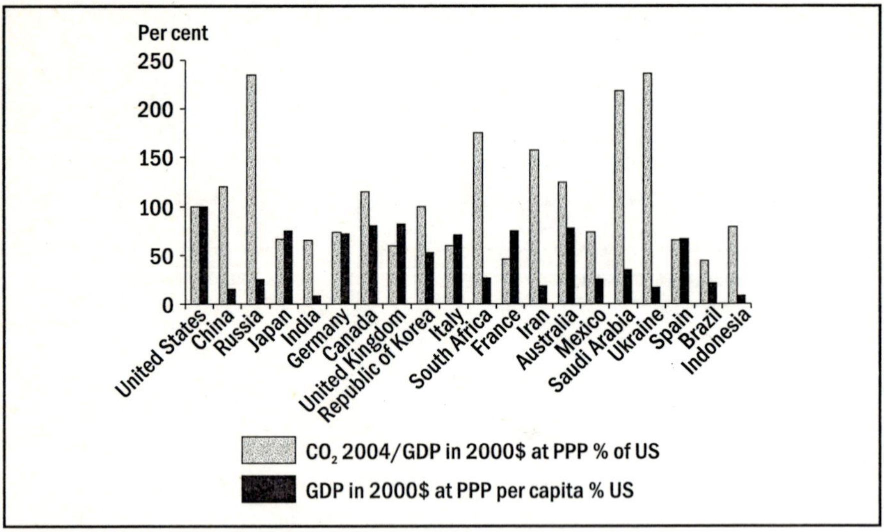

Figure 7 Carbon dioxide intensities of fossil fuels and gross domestic product per capita of the world's 20 largest economies

Note GDP – gross domestic product; PPP – power purchase parity

Source Bacon and Bhattacharya (2007)

Myth 3: India is unconcerned about its vulnerability to the impacts of climate change

Another common belief in developed countries is that India's alleged inaction on curbing GHG emissions stems from an irresponsible mindset among its policy-makers that India need not fear the ravages of climate change. There is no basis for this belief.

India has been highly vulnerable to climate variability (floods, droughts, cyclones, ocean surges) for millennia. Not just the present-day policy-makers, but the erstwhile British administrators, and before them, successive Indian dynasties, over the centuries, grappled with the impacts of the unpredictable climate. For many decades, India has had major, publicly funded programmes to address both the direct impacts, or prevention and control, of climate risks. In addition, other major public programmes focused on the creation of infrastructure or poverty eradication have had, as a major objective, the reduction of vulnerability to climate risks. At present, India spends no less that 12% of its annual budget, or 2.63% of the GDP, on these programmes. In fact, this is more than India's annual defence expenditure! Figure 8 provides some recent data on the aggregate expenditures, while Figure 9 depicts the programme areas where the money was spent. The programme areas were identified on the basis that each included programme would have among its stated objectives the objective of reducing vulnerability to climate variability.

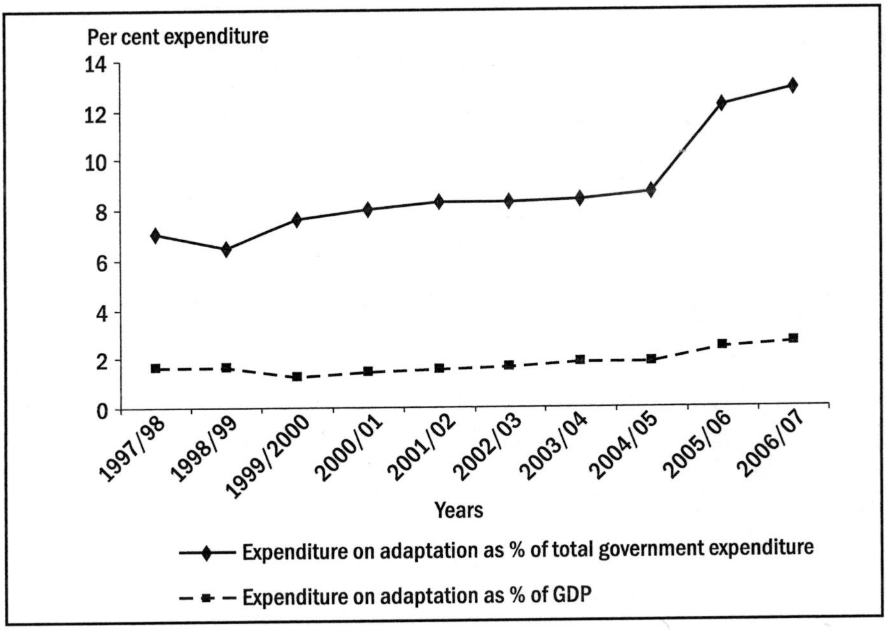

Figure 8 Annual expenditure incurred by India to address climate variability
Source GoI (various years)

Myth 4: India is an environmentally unsustainable economy

A curious perception is that India's relatively lower level of current and historical responsibility for climate change simply reflects its large-scale poverty, and as people become better off, they would quickly assume lifestyles prevalent in developed countries, but without the attendant environmental safeguards. The view does not account for a strong environmental ethic, specifically, a 'waste not' mindset, and deep reverence for nature and all living forms, which is deeply embedded in the culture of the people, and which remains unchanged with increased prosperity. Nor does it reflect awareness of the comprehensive policy and regulatory structure, besides publicly funded programmes, which are in place to address environmental concerns. Some international comparisons are given subsequently to illustrate, first, the fact that

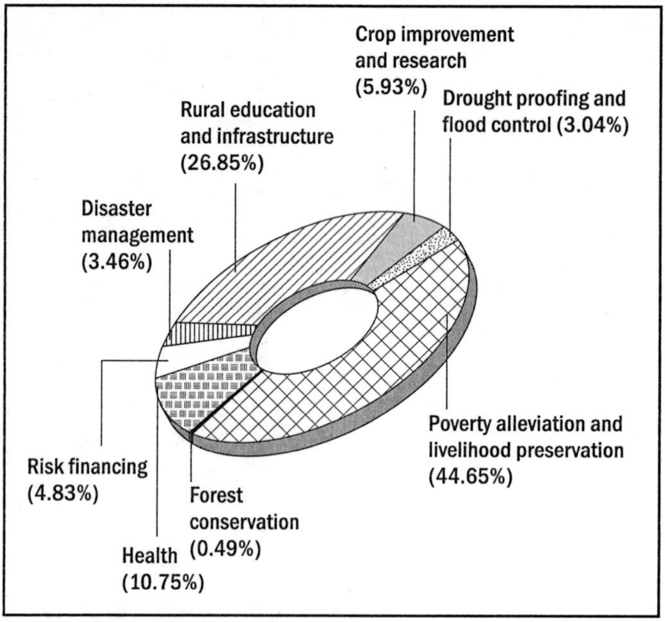

Figure 9 Proportion of critical components in total adaptation expenditure 2006/07
Source GoI (2007)

India has a strong environmental performance in terms of key sustainability parameters, which are indexed to eliminate the effects of incomes, and second, the outcome is traceable to environmental interventions of the government.

In case of India and China, CO_2 emissions from the food sector per kcal of food are an order of magnitude below that of the developed countries shown in Figure 10. A break-up of the emissions in each case into respective contributions of the food production and food processing (including packaging) components shows that in the case of developed countries, the latter dominates in terms of CO_2 emissions. Indians prefer fresh produce to processed food, and irrespective of economic status, buy fresh produce each day. Moreover, there is very little meat consumption (in terms of percentage of daily calories intake from meat), and this remains true even when people become richer.[8]

[8] Each calorie of food of animal origin requires approximately 10 times the energy required for producing the calorie equivalent of plant-based food.

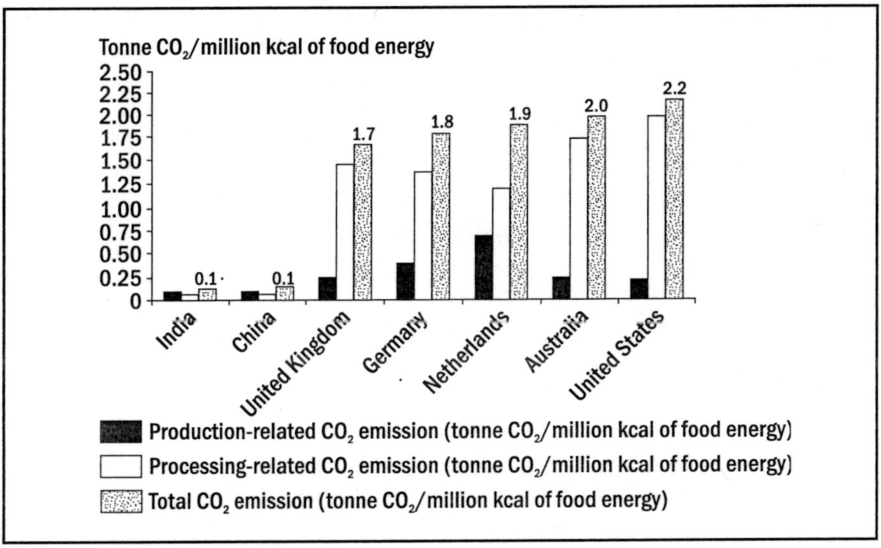

Figure 10 Carbon dioxide emissions from the food sector

Figure 11 presents a comparative data on the recycling rates of municipal waste for India and three developed countries.[9] India is

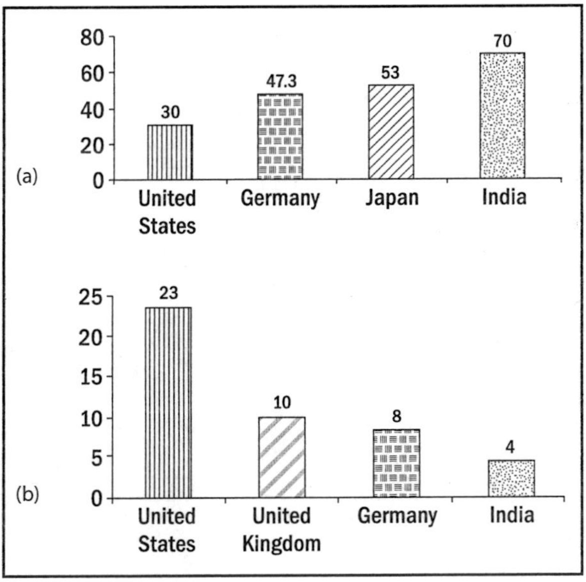

Figure 11 Recycling of solid waste and waste-related emissions:
(a) Average rate of recycling (%) (exclusive reuse)
(b) GHG emission from waste (gm/'000 GDP_{PPP})

Source TERI analysis

[9] Repair and reuse are not included in the data presented.

quite ahead of even Japan, the developed country with the most aggressive regulations to promote recycling. What accounts for the Indian performance? Very simply, India has a long, cultural tradition of recycling (as well as repair and reuse) – even wealthy households recycle everything possible – paper, metal, glass, and plastic. A well established network of non-formal sector recyclers visit every household at least once a month to buy the recyclables from the households. The recyclables actually comprise disposed of municipal garbage consisting mainly of kitchen waste, which can be composted rather than disposing it into landfills.

Another indicator of sustainability is CO_2 emissions from passenger transport.

Figure 12 presents a comparative account of data from India, EU-15, and the US on CO_2 emissions per passenger-kilometre of transport use. Again, India's emissions are less than one-seventh the emissions of EU-15, and just one-twelfth of that of the US. What accounts for this enormous difference? Notwithstanding the recent increase in the ownership of private vehicles, public and mass transport, that is, rail and bus, accounts for the major share of transport demand, including the annual incremental increase. Even with respect to automobiles, there is a strong cultural preference for fuel-efficient vehicles,[10] that is,

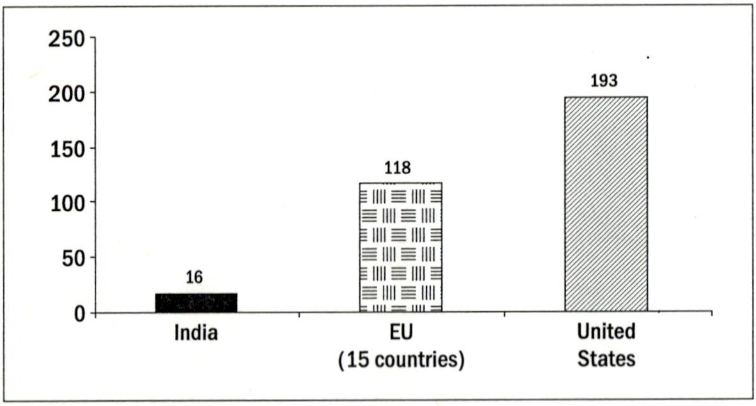

Figure 12 CO_2 emissions from passenger transport
Source TERI analysis (2005)

[10] This may be seen in advertisements for cars and two-wheelers. Even luxury vehicle manufacturers are careful to point out the fuel mileage of the vehicles.

cars and two-wheelers, and there is rapid increase in vehicles powered by natural gas, and more recently, especially in case of two-wheelers, electric vehicles are slowly being introduced.

There are numerous other dimensions of India's culture that complement sustainability. Most Indians are mostly vegetarian, signifying that only a small proportion of daily calorie intake is from foods of animal origin. Indians bathe twice a day, every day, but with a single bucket of water (25 litres). They switch-off all appliances promptly when not required.[11] They do not waste food. These habits do not change when Indians become richer. For example, they do not consume more meat. The *National Geographic's* Greendex, which evaluates a large set of developed and developing countries for environmental sustainability, has, in May 2009, ranked India as the world's most environmentally sustainable society.[12]

Sustainability is further depicted by a comparison of the so-called EKC, or Environmental Kuznets Curve, of India and some other countries. The EKC reflects a near-universal phenomenon that as countries grow, their environmental parameters first worsen and then improve, as higher incomes raise public environmental consciousness and also enable public resources to be diverted towards environmental management. A typical turning point for developed countries with respect to several sustainability parameters is an income level of approximately \$(PPP) 6000–7000. The concept of the EKC is illustrated in Figure 13.

Tables 2 and 3 present estimates of the turning points (in terms of per capita incomes) of the statistically estimated EKC curves from India and several other countries by various authors.

Table 2 presents the estimates of the EKC turning points for India and a set of 32 countries, which include both developed and other developing countries, for two key municipal waste water parameters in the receiving waters. The estimated turning points for India by various authors are much less than that for the set of 32 countries.

---

[11] Visitors to India observe that at night, there are virtually no lights left on in office and commercial buildings. This contrasts with what one observes in almost all developed countries.

[12] The first edition of the Greendex, in 2008, jointly placed India and Brazil as the world's most environmentally sustainable societies.

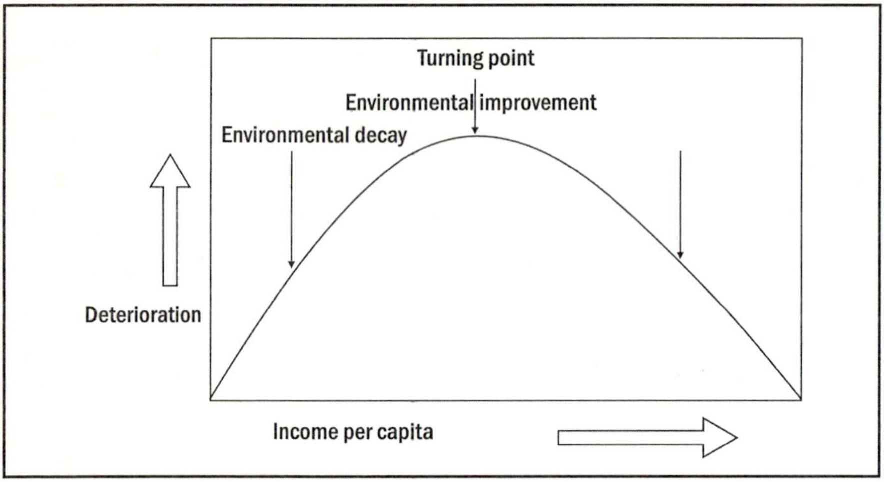

Figure 13 Illustration of the concept of Environmental Kuznets Curve, or EKC

Table 2 EKC estimates of turning points of India and a set of 32 countries (developed and developing) for wastewater treatment

	Countries	*BOD*	*COD*	*Wastewater*
MK ^	India	$523		
GK ^	Up to 32 countries	$7,623	$7,853	
CSB ^	India	$2,369		
NP ^^	India	$65 **		
Current study	India	$548 and $2,388*	$1,668 **	$3,150 (CI cities)
				$1,694 (CII cities)

MK = Mukherjee and Kathuria (2006)
GK = Grossman and Krueger (1995)
CSB = Chandra Sahu and Bali (2006)
NP = Narayanan and Palanivel (2003)
*N shaped, ** U shaped
^ in 1985 US $
^^ in 1995 US $
For the study by MK, a composite index of pollution including 63 environmental indicators has been used as the dependent variable
BOD – biological oxygen demand; COD – chemical oxygen demand; CI cities – Class I cities (five metros); CII cities – Class II cities, in India cities are classified in terms of size

Similarly, Table 3 gives the estimated EKC turning points for several key urban air quality parameters, that is, sulphur dioxide (SO_2), suspended particulate matter (SPM), and nitrogen oxides (NO_x). Once again, the estimated turning points in case of India are much lower than those for the other sets of countries.

Table 3 Air quality: comparisons of EKC turning points for India and several other country groups

	Countries	*SO_2*	*SPM*	*NO_x*
MK ^	India	$523		
CRB ^	11 OECD	$6,900	$7,300	$14,700
GK1 ^	Up to 32 countries	$4,107		
GK2 ^	Up to 32 countries	$4,053		
P1 ^		$3,000	$4,500	$5,500
P2 ^	30 developed and developing	$5,000		
SS ^	22 OECD and 8 developing	$10,700	$9,600	$21,800
S ^	31 countries	$3,670	$3,280	
Our estimates	India (ind, trans and res)**	$1,695	$1,640*	$1,707
		$957	$1,440*	$1,413*
		$1,752	$1,840	$1,770

MK= Mukherjee and Kathuria (2006)
CRB= Cole et al (1997)
GK1= Grossman and Krueger (1993)
GK2=Grossman and Krueger (1995)
P1= Panayotou (1995)
P2= Panayotou (1997)
SS= Selden and Song (1994)
S= Shafik and Bandhopadhya (1994)
^ in 1985 US $
* U shaped
**industrial, transport and residential sectors, respectively
for the study by MK, a composite index of pollution including 63 environmental indicators has been used as the dependent variable
Note 'Our estimates' refers to the TERI study, 2008

Similarly, Table 4 presents estimates of the EKC turning points for India and several other country groups with respect to energy intensity of the GDP. All developing countries (Bangladesh, India, and Sri Lanka) accomplished their turning points at much lower income levels than the three developed countries (Japan, Norway, and Switzerland). Of all countries, the turning point with respect to India was at the lowest per capita income level.[13]

Table 4 Energy intensity of gross domestic product: EKC turning points for India and several other countries

	Specification	*Shape*	*Turning point*	*Current income**
Bangladesh	Quadratic	EKC	$1,377	$1,827
India	Country	EKC	$501	$3,072
Japan	Quadratic	U	$22,675	$27,817
Netherlands	Linear	Monotonically decreasing	—	$29,078
Norway	Quadratic	EKC	$10,274	$36,849
Pakistan	Linear	Monotonically decreasing	—	$2,109
Sri Lanka	Quadratic	U	$4,092	$4,088
Sweden	Linear	Monotonically decreasing	—	$28,936
Switzerland	Quadratic	EKC	$26,122	$31,701
UK	Linear	Monotonically decreasing	—	$29,571

*GDP per capita at constant 2000 international $, PPP in 2005

Note Where the EKC curve is stated to be 'monotonically decreasing' sufficient past data has not been available to estimate the turning points.

Source TERI study (2008)

Myth 5: Never mind the past, in the future, India's GHG emissions will grow uncontrollably

The fact of India's recent high GDP growth rate (approximately, 7%–8% per annum for the last decade, prior to the global economic downturn) makes some people in developed countries nervous that over a generation, this would result in extremely high levels of GHG emissions. Some modelling studies carried out by institutions in developed countries, which essentially extrapolated the experience of the developed countries in an earlier era, indeed seem to suggest that this might be the case. Essentially, these studies neglected the fact that

[13] Sufficient past data was not available to estimate the turning points with respect to the other countries studied (the Netherlands, Pakistan, Sweden, and UK).

the rapid decline in India's energy intensity (and all modelling results show that the energy intensity of the economy will continue to decline), resulting in a GDP growth rate of 8% per annum being accomplished at no more than 3.7% increase in energy use, will significantly moderate the growth of energy use in India.

A Computable General Equilibrium, or CGE, modelling study by the NCAER (National Council of Applied Economic Research), New Delhi, which accounts for the observed increasing energy efficiency (as well as increased efficiency of production sectors generally) provides some interesting results. First, in the period till 2030/31, the trend depicting rate of GDP growth remains in the range of 8.5%–9% per annum (Figure 14), assuming that recently observed annual energy efficiency increase[14] of 3% and the overall efficiency of resource use[15] continue at a rate of 1.5% per annum. At the same time, under these assumptions, the energy intensity of the GDP continues to decline, from 0.11 kilogram oil equivalent (kgOE)/$GDP[16] in 2003/04 to 0.04 kgOE/$GDP in 2030/31 (Figure 15).

Further, far from a runaway growth of CO_2 emissions,[17] the Indian per capita CO_2 emissions are expected to grow only modestly, from 1.0 tonne per capita in 2004/05 to 2.77 tonnes per capita[18] in 2030/31. This may be compared with the 2005 global average CO_2 emissions of 4.22 tonnes per capita in 2005 (Figure 16). In other words, even after a whole generation of high growth, India's per capita GHG emissions would be well below today's global average.

Lest this result be felt to be implausible, it may be added that a recent World Bank study, *India: low carbon growth*, using a different model, has arrived at virtually identical results.[19]

---

[14] Called Autonomous Energy Efficiency Improvement, or AEEI, parameter.

[15] Called Total Factor Productivity Growth, or TFPG, parameter.

[16] The model uses the actual calorific values of Indian fuels, and accordingly, the energy intensity given by the model is somewhat lower than that given by the IEA, which employs European norms.

[17] Due to fossil fuels and industrial emissions.

[18] While India's population is large in absolute numbers, that is, 1123 million in 2007, the rate of population growth is approximately 1.4% per annum, a rate typical of Latin American countries, which are much richer. The CGE model uses population growth projections of India's Registrar General of Census.

[19] Both the NCAER and World Bank modelling studies may be presented to a wider audience shortly.

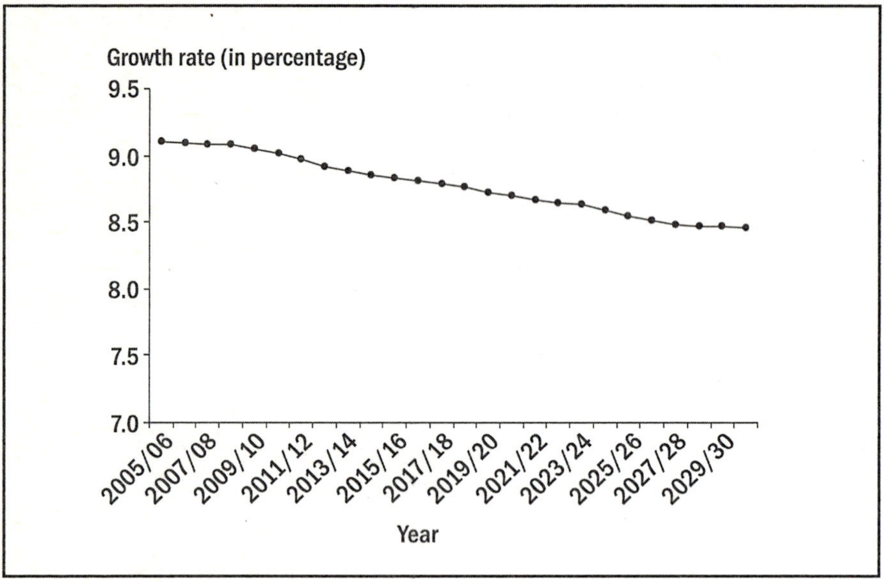

Figure 14 Trend of GDP growth rate till 2030: Computable General Equilibrium model simulation

Source NCAER Study (2009)

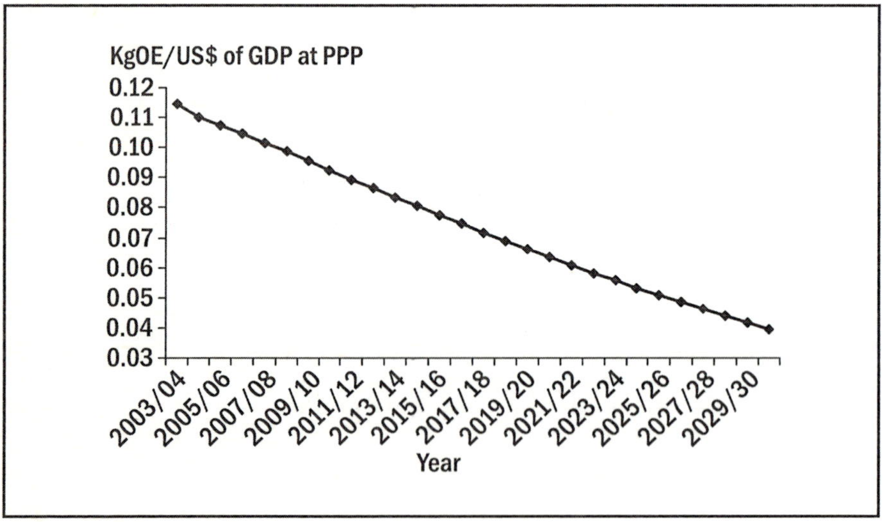

Figure 15 Trends in energy intensity of GDP till 2030: Computable General Equilibrium Model simulation

Source NCAER Study (2009)

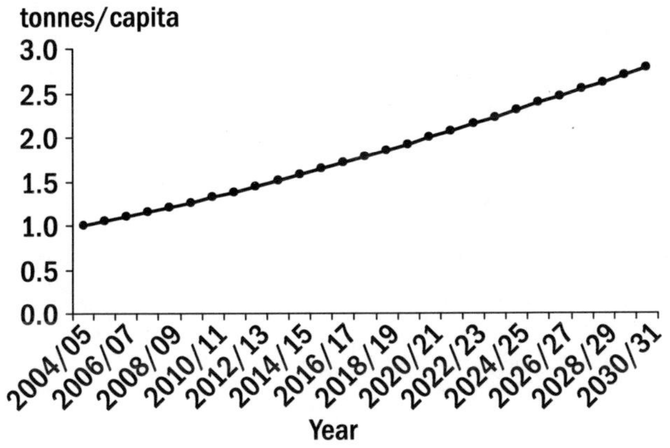

Figure 16 Trends in per-capita CO_2 emissions till 2030: Computable General Equilibrium Model simulation
Source NCAER Study (2009)

Myth 6: What is all the fuss about? Reducing GHG emissions pays for itself or, at best, is low cost

There is a widespread view that developing countries in general, and India in particular, are under a serious misconception that reducing GHG emissions or at least limiting their growth rates will adversely impact their economies. What does actual empirical research say on this issue?

Figure 17 presents the result of the simulations of the MARKAL[20] model to evaluate the potential and costs of GHG mitigation in five sectors, that is, SMEs (small and medium enterprises), (conventional) power, renewable energy, residential and commercial, and bio-diesel, for the period 2016–36. The maximum GHG reduction from the baseline, which are feasible, is just 9.7%. The incremental investment cost of doing so is $2.3 trillion. The only way India could, on its own, meet this additional increment cost is by diverting development resources from schools, hospitals, village roads, railways, mass transport, immunization, child nutrition, maternal mortality, and so on.

[20] MARKAL, or Market Allocation, is a bottom-up engineering–economic linear programming model developed by the International Energy Agency, Paris, and is widely used worldwide for identifying optimal energy technology choices and evaluating costs of greenhouse gas mitigation.

What of the increased energy costs, if any, of the 9.7% reduction from baseline accomplished? Figure 18 gives the results of the same simulation in this respect. About 3% of the GHG reduction from the baseline (far left of graph) may be accomplished with reduced energy costs. For the remainder of 9.7% GHG mitigation from baseline in 2036, the discounted economic cost is $180 billion!

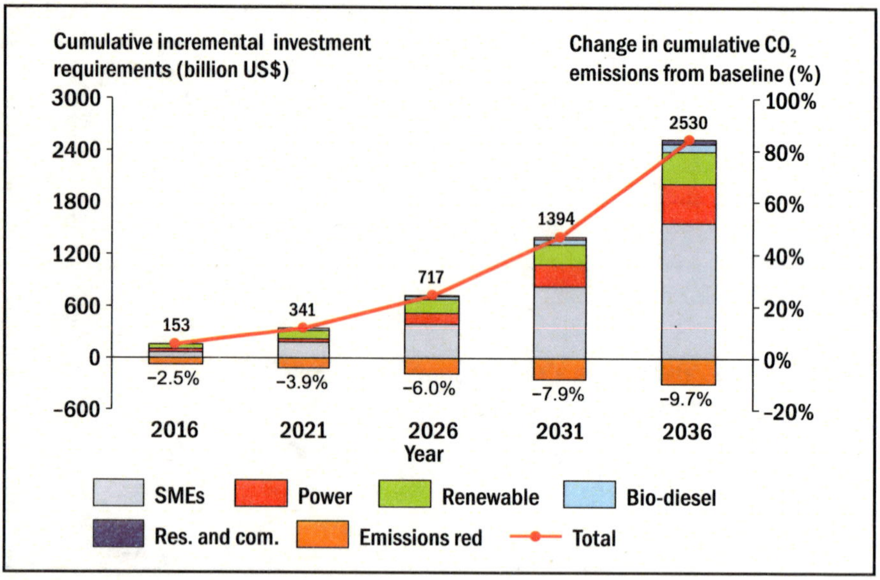

Figure 17 MARKAL simulations of costs of GHG mitigation in five sectors
Source TERI study (2005)

The diversion of investment resources to GHG mitigation from other sectors has adverse implications for economic growth as well as for poverty levels. Figure 19 presents the results of the simulation of the CGE model (referred to previously) with respect to the imposition of a revenue positive carbon tax, generally recommended in the environmental economics literature as an efficient policy instrument, on GDP growth, for the period 2004/5 to 2030/31. There is GDP reduction of 8.1% from the baseline with a carbon tax of $80 per tonne of CO_2. The cumulative GDP loss (undiscounted) with the CO_2 revenue positive carbon tax at the $80 per tonne level is $5.7 trillion! (Figure 19).

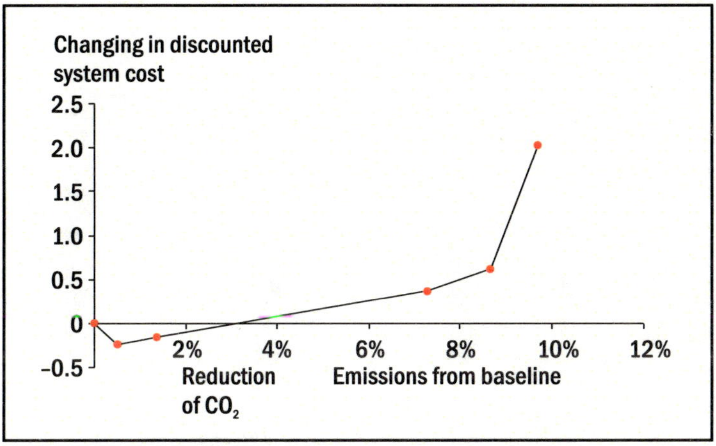

Figure 18 Discounted incremental energy costs of MARKAL simulations of greenhouse gas mitigation in five sectors
Source TERI study (2005)

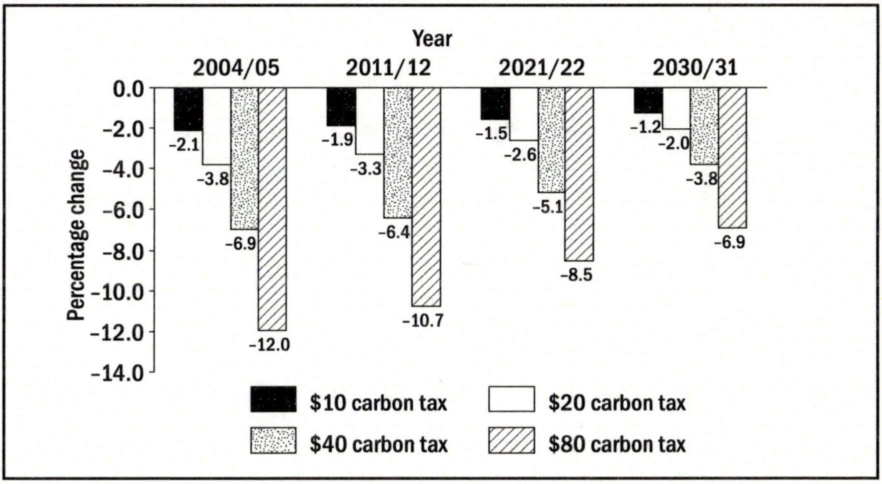

Figure 19 Impact of a revenue positive carbon tax on gross domestic product
Source NCAER Study (2009).

How much GHG reduction is accomplished by this level of tax? Just 2.6 tonnes per capita with a dollar carbon tax from 2.8 tonnes per capita without tax in 2030/31 (Figure 20).

Suppose, however, the carbon tax were revenue neutral, an option favoured by many public finance economists, that is, the carbon tax revenue was balanced out by an equivalent reduction in direct taxes.

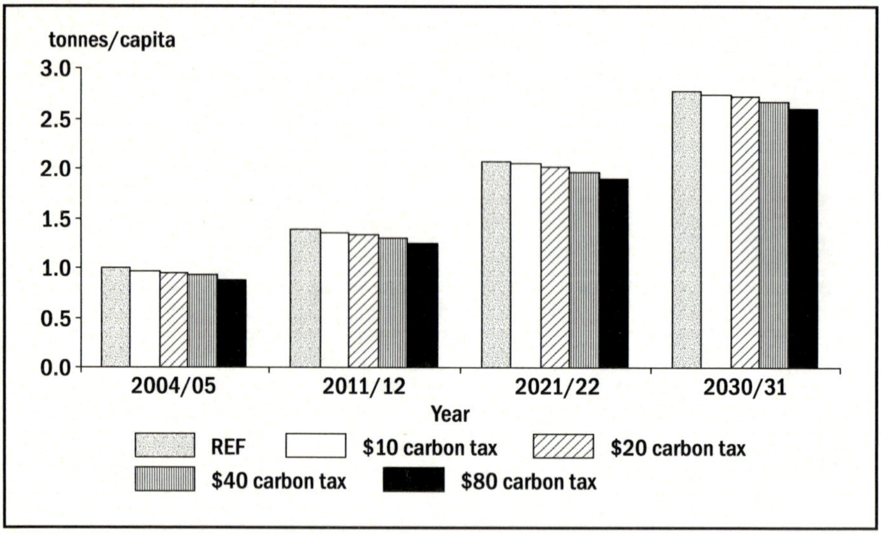

Figure 20 Greenhouse gas reductions per capita from the imposition of a revenue positive carbon tax
Source MoEF (2009)

In that case, the GDP reduction with a $80 per tonne CO_2 carbon tax is marginally reduced to 6.9% in 2030/31. The (undiscounted) cumulative GDP losses at the $80 per tonne CO_2 tax level are also marginally reduced to $5.1 trillion (Figure 21). The impact, at this tax level, on GHG emissions per capita in 2030/31 is unchanged at a reduction of 2.6 tonnes per capita.

Clearly, confronted with these results, no Indian policy-maker can view the prospect of uncompensated aggressive action to mitigate GHG, beyond the current policy and regulatory regime with equanimity. The World Bank's study, *India: low carbon growth*, has also come to the conclusion that any GHG mitigation actions by India, beyond the current development plans, would seriously impact GDP growth and poverty alleviation.

The way forward – India's perspectives

Given this background, let us now look at what India's proposals are, so far, for moving forward on the key issues in the global climate

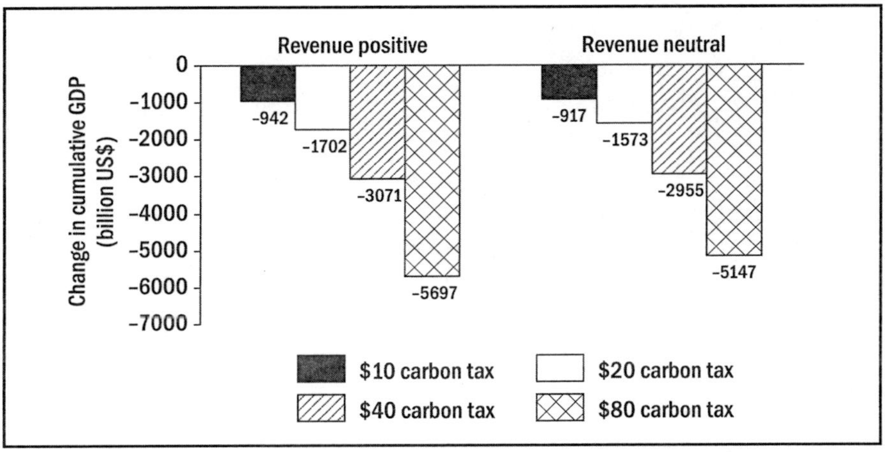

Figure 21 Cumulative (undiscounted) gross domestic product losses with carbon tax
Source MoEF (2009)

change agenda.[21] For the sake of convenience, India's approach can be looked at in terms of 'building blocks' and some other key elements of the Bali Action Plan, or BAP.

GHG mitigation

In India's view, developed countries need to commit to deep, long-term, legally binding GHG reductions, consistent with the espoused level and time frame of GHG stabilization. GHG mitigation actions by developing countries must be enabled and compensated for by financial and technology transfers from developed countries. There can be no differentiation of 'key developing countries' from other developing countries.

Figure 1 revealed the extent of historical responsibility of several developed countries. Without a meaningful discharge of this responsibility, based on an explicit recognition that all humans have equal rights to the atmospheric resource, besides termination of their current unsustainable emissions levels, no climate regime would be seen to be fair or to solve the problem. On the other hand, significant mitigation actions by developing countries beyond the efforts that they

[21] In fact, India's approach to the global regime is entirely consistent with the United Nations Framework Convention on Climate Change (UNFCCC), the Kyoto Protocol, and the Bali Action Plan. Most elements also have the endorsement of the group of G77 and China.

are even now making will lead to major diversion of their resources away from development and poverty eradication, unless these are adequately compensated and the necessary technology is provided at low cost. The so-called 'key developing countries', again as shown in Figure 1, have, in fact, provided all the environmental space used by all the developed countries together. Their development challenges are typical of other developing countries. They cannot be penalized by abridging their development on the argument of 'size', when they have, in fact, enabled the developed countries to industrialize, and their energy or CO_2 intensities are no different from those of developed countries in general.

Sectoral targets

In India's view, sectoral emission targets by developing countries in line with externally imposed norms will only create a market opening for technologies closely held by a few developed country firms; these may be inappropriate to the situations in developing countries and will involve high, uncompensated technology costs due to cartelization by the suppliers. In any event, no sector can be defined unambiguously, after taking into account vintages, technology pathways, differences in raw materials, finished products, transport linkages, and so on, and as such, sectoral norms are in practical terms a non-sequitur.

Externally imposed sectoral targets are an inefficient and impractical means of GHG mitigation. They are primarily intended to gain market access, and in any event are not permissible under the BAP (para 1(b)(iv)), which speaks only of enhancing implementation of para 4.1(c) of the United Nations Framework Convention on Climate Change (UNFCCC), which, in turn, speaks only of promotion and cooperation in development, application and diffusion of technologies, and practices and processes for the mitigation of GHG in relevant sectors.

Nationally appropriate mitigation actions

The BAP requires developing countries to formulate and implement Nationally Appropriate Mitigation Actions (NAMA), supported and enabled by finance, technology, and capacity building that are 'monitorable, verifiable, and reportable, or MRV. In India's view 'nationally appropriate' signifies that the plans must be prepared by the countries themselves, without external dictation or 'adjustment of ambition'.[22]

Mitigation actions by developing countries should proceed on this basis, and the actions themselves, provided these are supported by finance, technology, and capacity building, as well as the support provided is accountable in the MRV sense. Actions carried out by developing countries on their own, without support, cannot be subject to MRV accountability.

Financing

In India's view, financial support for NAMAs of developing countries are not 'aid', but a discharge of responsibility by developed countries, scaled both by their historical responsibilities for climate change and the capabilities they have acquired thereby.

The 'aid' paradigm involves discretion by 'donors' as regards volume of resources, the purposes for which they may be used, the sources of skills, equipment, and technologies to which the aid may be applied, the countries and organizations within them that may receive the 'aid', and the institutions through which the 'aid' may be provided.

The responsibility-based approach, on the other hand, signifies that the resources must be 'new and additional' (that is, not diverted from 'aid'), assessed and not discretionary, and be administered by a financial mechanism answerable to the parties, with a unique governance structure. Financial resources amounting to 0.5%–1.0% of aggregate GDP of developed countries are necessary to address GHG mitigation in developing countries.

Technology

In India's view, technology is the key to addressing climate change in both the aspects of mitigation as well as adaptation. It comprises three elements. First, a global effort on research and development (R&D),

---

[22] It is sometimes argued by developed country policy-makers that without such external oversight, a developing country's supported NAMA actions may be negated by actions outside the supported NAMAs. This demand is unreasonable on three counts. First, a country's public policies, legislations, regulations, and budgets are always in the public domain. Second, it would be irrational for a developing country to deviate from a baseline action to a economically suboptimal action outside the supported NAMAs. Third, would any developed country accept such external oversight in respect of its own policies, legislation, regulations, and budget?

including adaptive R&D to enable deployment of available technologies in developing countries as well as to develop new, cost-effective clean technologies. This will involve significantly stepped-up public as well as corporate financing on R&D in developed countries. In addition, R&D efforts should involve partnerships between institutions in developed and developing countries, with sharing of IPRs (intellectual property rights), and these collaborative R&D efforts may be financed through the Financial Mechanism of the UNFCCC.

Second, available and new clean technologies must be available to developing countries for their climate change actions on non-commercial terms. This signifies that (1) the existing social contract on IPRs must be tempered with to balance the rewards to the innovator while addressing the global imperative of saving the climate, involving negotiated or regulated,[23] rather than monopolistic, license fees, as indeed the international community has agreed in case of drugs needed for HIV/AIDS and (2) that where particular clean technologies are of wide applicability in developing countries, the Financial Mechanism may purchase the technology rights for their use in the context of their climate change actions.

Third, a network of regional technology innovation centres should be set up in developing countries to catalyse collaborative R&D, provide reliable information on available technologies, their costs, and performance, and enable capacity building on deployment of clean technologies and their further innovation.

Adaptation

Adaptation has been the Cinderella of the climate change regime. It is clear from various studies, as well as country experience (for example, India's experience discussed previously) that the resource and technology needs for adaptation are of the same order as those for mitigation. 'Mainstreaming' adaptation actions into development programmes must not involve a diversion of development resources to adaptation, whether the country's own or provided externally. All

---

[23] There is a widespread misconception that developing countries are asking for technology to be provided 'free'. A regulated license would be no more free nor provide insufficient incentive to the innovator (entrepreneur) than regulated tariffs for any natural monopoly, say, electricity.

vulnerable regions, and not only the least-developed or small-islands must receive adaptation funding from the Financial Mechanism, as indeed the UNFCCC and Bali Action Plan contemplate.

Sustainable production and consumption

High per capita GHG emissions in developed countries are the inevitable outcome of unsustainable lifestyles, comprising unsustainable patterns of production and consumption. These need to be addressed in the future climate change arrangements, and it must be recognized that, on the one hand, human well being is not conditional on unsustainable lifestyles,[24] and on other, the argument that the present lifestyles of certain countries are sacrosanct, is untenable.

Conclusions

The conclusions can be summed up in following four points.

1. First, that India was not in the past, is not now, nor likely in the future, a part of the climate problem. Rather, it has always been responsible in terms of energy use and environmental concerns.
2. Second, that India's (and all developing countries') concerns about economic growth and poverty eradication are legitimate, and must be fully respected in any global climate regime, as indeed stated unequivocally in the UNFCCC and the BAP.
3. Third, that the cause of climate change, and one which is continuing, is the unsustainable emissions of developed countries. They have to take leadership to drastically reduce their emissions, and this will involve modification of their lifestyles.
4. Finally, that the proposals made by India (and other developing countries) with respect to the future climate regime are constructive, and must be given serious consideration in any discussions on global climate action.

---

[24] For example, an assertion that commuting to work every day in an SUV with single occupancy, facing traffic congestion arising from everybody else doing the same, as opposed to commuting by safe and efficient mass transport, enhances well-being, is on the rise.

References

Bacon R and Bhattacharya S. 2007
Growth and CO2 Emissions – how do different countries fare?
Washington, DC: World Bank

BEE (Bureau of Energy Efficiency). 2007
Personal communication

FICCI (Federation of Indian Chambers of Commerce & Industry). 2007
FICCI Climate Change Task Force Report 2007
New Delhi: FICCI

GoI (Government of India). 2007
Data Union Budget documents 2007/08
New Delhi: GoI

GoI (Government of India). Various years.
Data Union Budget documents
New Delhi: GoI

IEA (International Energy Agency). 2006
Database 2006
Details available at < http://www.iea.org/stats/index.asp>, last accessed on 12 November 2006

MoEF (Ministry of Environment and Forests). 2009
India's GHG Emissions Profile: results of five climate modeling studies
Details available at <moef.nic.in/downloads/home/GHG-presentation.pdf>, last accessed on 30 December 2009

UNDP (United Nations Development Programme). 2008
Human Development Report 2007/08
New York: UNDP

World Bank. 2009a
World Development Report 2009
Washington DC: World Bank

World Bank. 2009b
World Development Indicators Database
Details available at <http://web.worldbank.org/WBSITE/EXTERNAL/DATASTATISTICS/0,,contentMDK:21725423~pagePK:64133150~piPK:64133175~theSitePK: 239419,00.html>, last accessed on 15 October 2009

9

Emission reduction targets: a Pandora's Box in climate negotiations

Leena Srivastava and Neha Pahuja*

The Bali Action Plan (BAP) adopted at the 13th session of the Conference of the Parties (COP-13) of the United Nations Framework Convention on Climate Change (UNFCCC) laid four building blocks – mitigation, adaptation, technology, and finance – to enhance implementation of the climate change regime.[1] Since then, the mitigation debate has revolved around issues involving the setting up of a long-term global emission reduction goal to achieve the ultimate objective of the Convention, that is, stabilization of greenhouse gas (GHG) concentrations in the atmosphere at a level that would prevent dangerous anthropogenic interference with the climate system.[2] This would involve the distribution of this goal amongst individual countries and in doing so, defining mitigation commitments and targets of the developed countries and voluntary mitigation actions of the developing countries; and means of implementation of these targets in terms of relevance of domestic measures and carbon markets for developed countries and support in the form of technology, finance and capacity building from developed to developing countries. Much of the debate, however, is rooted in the differential recognition and treatment of developed and developing countries in the Convention and the Kyoto Protocol (KP) owing to common but differentiated

---

* Leena Srivastava, Executive Director, TERI, and Neha Pahuja, Research Associate, TERI

1 UNFCCC (2007), Decision 1/CP.13, Bali Action Plan

2 UNFCCC (1992), United Nations Framework Convention On Climate Change, Article 2

responsibilities and respective capabilities and socio-economic conditions of the Parties.[3]

International deliberation in Bali in 2007 launched a comprehensive process, with two tracks—the Ad hoc Working Group on further commitments of Annex I Parties under the KP (AWG-KP) and the Ad Hoc Working Group on Long-term Cooperative Action under the Convention (AWG-LCA), stated to culminate in Copenhagen at the 15th session of COP of the UNFCCC. It was believed that the Copenhagen outcome would resolve the mitigation debate by moving a step further in bridging the developed and developing country divide. However, the outcome of this session, the Copenhagen Accord[4], could not resolve the complex mitigation debate and left the world further divided, now with a number of smaller groups being formed. This paper dwells on the impasse between the developed and the developing countries, especially on the mitigation debate in the international climate change regime, and highlights the complexities involved in arriving at a consensus, given the economic and political interests of each country. The paper, thus, focuses on issues involved in the international debate on mitigation. The way forward would be to resolve these complex issues to enable the required mitigation action to combat climate change.

Revisiting the Kyoto Protocol

One of the important issues that preceded the Copenhagen session and remained unresolved is whether the KP will continue to exist with a further strengthened form or a new instrument will find its invent. The issue holds much political significance because it is the only legal instrument that exists in the climate change regime. While developing countries argue for more stringent reduction targets from the developed countries keeping the KP intact, developed countries argue for reducing the groupings within the KP, thereby facilitating the inclusion of the US and fast developing countries in the group that accepts legally binding emissions reduction targets.

The KP sets binding targets for 37 industrialized countries for reducing GHG emissions, which amount to an average of

[3] UNFCCC (1992), United Nations Framework Convention On Climate Change, Article 3.1

[4] The Copenhagen Accord has only been 'taken note of' and is not a COP decision

approximately 5% against the 1990 levels over the five-year commitment period of 2008–12.[5] The KP had emissions reduction targets on these 37 nations, recognizing that the developed countries were principally responsible for the high levels of GHG emissions that have resulted from more than 150 years of industrial activity. They, therefore, have a larger burden to reduce emissions on the basis of the principle of common but differentiated responsibilities (CBDR). The KP, often seen as an important first step towards a truly global emission reduction regime to be able to stabilize GHG emissions and as an ambitious outcome of a decade-long UN process with almost all nations being signatories to it, however, faced a near fatal rejection even before its coming into force with the then world's largest emitter, the US, failing to ratify it. The US, from the very beginning, demanded binding emissions reduction targets even for the developing nations, as failure to do so would seriously impact its economy. It was at this stage that the international community began the debate on the design of what could be an even more comprehensive successor to the KP, which could embrace the US and other large emitters from the developing world.

With rapid economic growth in fast developing countries like China and India and the subsequent change in global emissions profile (increased emission flows from developing countries), the developed countries contended even more strongly and cohesively for the KP being insufficient, as it does not legally oblige these fast developing countries to reduce their emissions. An analysis (Figure 1) of the global emissions profile shows that while the aggregate share of Non-Annex I emissions was miniscule in 1950, it has now substantially increased. Recent emissions data shows that in 2006, aggregate share of Non-Annex I countries was more than 50% of the total global emissions. Further, analysis focusing on the list of top 10 emitting countries during 1950 (an historic benchmark) (Table 1), 1990 (the baseline year in KP), 1997 (adoption of KP), 2005 (KP entered into force) and 2006 (most recent data signifying China superseding the US) shows the inclusion of increasing number of fast developing countries in the list. Hence, emphasizing that though the developing countries had a meagre contribution to the historical global emissions,

[5] UNFCCC (1998), Kyoto Protocol to the United Nations Framework Convention On Climate Change, Article 3.1

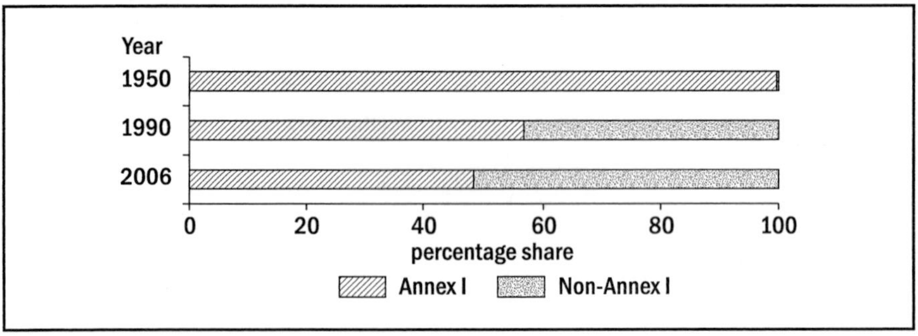

Figure 1 Change in global emissions profile
Data source WRI (2010)

they are now rapidly emerging as leading emitters, especially China, with about 11% of the global emissions in 1990 almost doubling to 21.8% in 2006, and now becoming the world's largest emitter.

Table 1 Top 10 emitters

1950	1990	1997	2005	2006
US (42.3)	US (23.3)	US (24.2)	US (21.3)	China (21.8)
EU-27 (30.1)	EU-27 (19.8)	EU-27 (17.5)	China (20.3)	US (20.3)
Germany (8.7)	China (11.0)	China (14.6)	EU-27 (14.9)	EU-27 (14.5)
UK (8.5)	Russia Fed (10.5)	Russia Fed (6.4)	Russia Fed (5.6)	Russia Fed (5.7)
Russia Fed (7.1)	Japan (5.3)	Japan (5.3)	Japan (4.6)	India (4.7)
France (3.4)	Germany (4.6)	India (4)	India (4.5)	Japan (4.4)
Canada (2.6)	Ukraine (3.3)	Germany (3.9)	Germany (3)	Germany (3)
Ukraine (2.0)	India (3)	UK (2.3)	Canada (2)	Canada (1.9)
Poland (1.9)	UK (2.7)	Canada (2.2)	UK (2)	UK (1.9)
Japan (1.7)	Canada (2.1)	S Korea (2)	S Korea (1.8)	S Korea (1.8)

Note The figures in brackets indicate the percentage share of the country in global emissions and Non-Annex I countries are indicated in black.
Data source WRI (2010)

The analysis of the emission trends of major Annex I countries (Figure 2) illustrates that while emissions of the US have substantially increased 1950 onwards, for other countries it has either increased a little, decreased or stabilized. It is for the same reason (that is, the US being largest emitter and its emissions increasing substantially over time) that other developed countries demand its presence in any legally

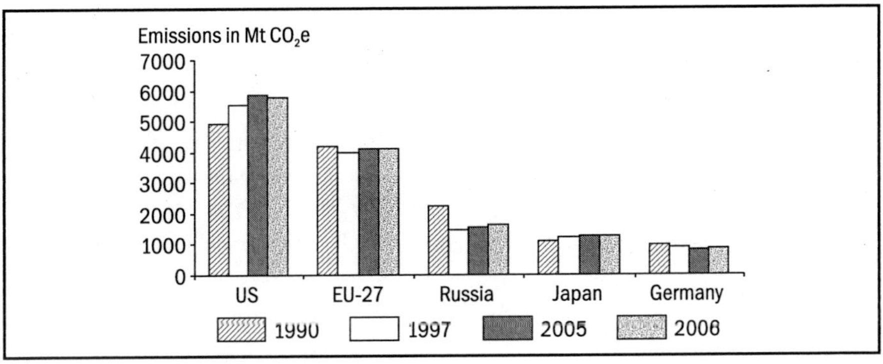

Figure 2 Emissions in major Annex I countries
Note $MtCO_{2e}$ – million tonnes of CO_2 equivalent
Data source WRI (2010)

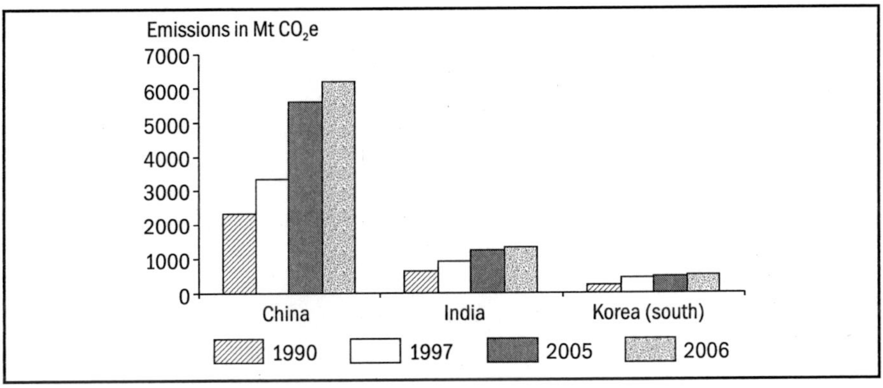

Figure 3 Emissions in major Non-Annex I countries
Note $MtCO_{2e}$ – million tonnes of CO_2 equivalent
Data source WRI (2010)

binding instrument to curb global emissions under an international climate change regime.

Similar analysis of the emissions trends of major Non-Annex I countries (Figure 3) illustrates that most of these countries have shown increasing emissions. However, the emissions of China are evidently much higher than that of other countries and the 2006 emissions are almost three times that in 1990. It is for the same reason (that is, rapid growth in emissions and the substantial share in global emissions) that China is being pressurized by the developed nations to accept binding emissions reduction target. Also, the fact that China has now gradually surpassed the US and the EU to become the world's largest contributor of the present emissions flow (Figure 4) has led the international community to the idea of revisiting the KP.

Hence, the changing global emissions profile has sprouted lot of proposals on regrouping of countries, graduation of countries, scheduled approach, targets for all, and so on. However, a way forward towards the post-2012 scenario will be to resolve this contention amongst all countries—whether the global conditions have changed to the extent that requires regrouping or distribution of countries. This can be achieved with further deliberations on the criteria that should lead to this regrouping. Also, the principles of the Convention that formed the basis of the KP should be kept intact, as it gives due consideration to socio-economic conditions of a country. Therefore, a holistic approach will be required, which considers socio-economic conditions along with the changing emissions profile to be able to reach a larger global acceptance. These criteria will provide the essential architecture for future international agreement on climate change.

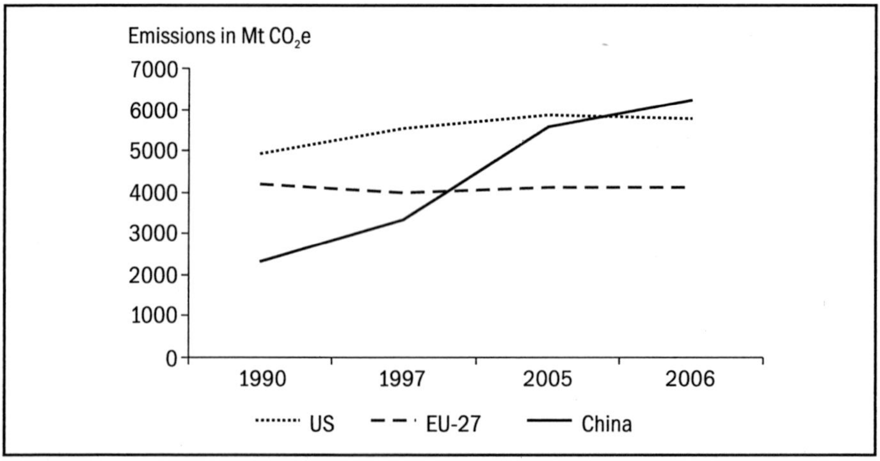

Figure 4 Comparison of emissions: US, EU and China

Note $MtCO_{2e}$ – million tonnes of CO_2 equivalent

Data source WRI (2010)

Binding global emission reduction target: what does it entail?

Another important issue that the world leaders, negotiators, environmentalists, and scientists could not resolve is that of an agreed upon global emissions reduction target. The key issues involved in setting up of a binding global emission reduction target are the form the target should be in, the scale of the global target, and the distribution of targets among individual countries. Resolving these issues would be a crucial precondition for a successful and ambitious international framework or agreement on climate change.

The Intergovernmental Panel on Climate Change (IPCC 2007) estimates Annex I to reduce emissions by 25%–40% from the 1990 level and Non-Annex I countries to observe substantial deviation from the baseline in regions like Latin America, Middle East, East Asia and Centrally Planned Asia by 2020. Also, it estimates Annex I countries to reduce by 80%–95% from the 1990 level and all regions in Non-Annex I to observe substantial deviation from the baseline by 2050 for the 450 ppm CO_2-equivalent scenario. Hence, the negotiations so far centred around the understanding that by the end of the KP's first commitment period in 2012, an international framework would need to be negotiated and ratified, which can deliver the required stringent emission reductions as indicated by the IPCC. This was emphasized in Para 1(a) of the BAP, which calls for setting up of a long-term global goal/target for emission reductions as part of the shared vision for a long-term cooperative action.[6] Since then, a variety of goals relating to emissions reductions that would lead to stabilization of GHG concentrations with varying degrees of specificity have been proposed.[7] For instance, some propose stabilization of GHG concentrations in the atmosphere at around 450 ppm carbon dioxide equivalent (CO_2 eq), while others suggest 350 ppm as most appropriate, translating into a temperature limit of 2 °C and 1.5 °C respectively.

However, a recent paper by Winkler, Vorster, and Marquard (2009) points out that implicit in the IPCC numbers of the global range and the Annex I target is a formula that implies that Non-Annex I countries will be responsible for undertaking emissions reduction equivalent to the difference between a global emissions target and the quantified ranges for Annex I (specific ranges for Non-Annex I have not been quantified by the IPCC), given the fact that sharing the burden for mitigation is essentially zero sum, that is, if one does less, the other has to do more. The same paper also cites den Elzen and Höhne (2008) for their analysis on the implicit Non-Annex I targets, which range from 15%-30% for 450 ppm. The same range was discussed in the AWG LCA technical

[6] UNFCCC (2007), Decision 1/CP.13, Bali Action Plan, Para 1(a)

[7] UNFCCC (2009), AWG-LCA negotiating text, FCCC/AWGLCA/2009/8

briefing[8] and the Council of the European Union also concluded the need for a deviation from the baseline by developing countries, ranging from 15%–30% below business as usual.[9] The TERI analysis (Table 2) based on emissions data from WRI CAIT indicates that only a global emissions reduction target of 50% from the 1990 levels combined with Annex I Parties' reduction of 90%–100% from the 1990 levels can allow Non-Annex I Parties to increase their emissions, and that too marginally. However, higher global emission reduction combined with not so ambitious Annex I emissions reduction targets would mean Non-Annex I having to contribute more than Annex I. This has great political significance, as it significantly deviates from the principles of the Convention, especially CBDR.[10]

Table 2 Implied emission changes for Non-Annex I countries

Annex I emission reduction from 1990 levels (%)	90	80	70	60	50	40
Global emission reduction scenario	Non Annex I emissions change					
Scenario 1 (50% reduction of 1990 levels by 2050)	12	-4	-19	-35	-50	-65
Scenario 2 (70% reduction of 1990 levels by 2050)	-39	-55	-70	-85	-101	-116
Scenario 3 (80% reduction of 1990 levels by 2050)	-65	-80	-95	-111	-126	-142
Scenario 4 (85% reduction of 1990 levels by 2050)	-77	-93	-108	-123	-139	-154

Data source WRI (2009)

White box signifies increase in emissions, grey box signifies emission reductions but less than that of Annex I, black box indicates emission reductions more than that of Annex I

Further analysis shows (Figure 5) that a 50% global emission reduction from 1990 levels by 2050 combined with an 80% emission reduction from 1990 levels by 2050 will still not achieve global convergence of per capita emissions. The analysis predicts even if most of the emission reduction in Annex I takes place through domestic

---

[8] http://unfccc.int/files/kyoto_protocol/application/pdf/emission_reductions_for_stabilizing.pdf

[9] Commission of the European Communities, 2009, Part I of extensive background information and analysis, commission staff working document accompanying the communication from the commission to the European Parliament, the European Economic and Social Committee and the Committee of the Regions: towards a comprehensive climate change agreement in Copenhagen, available at http://ec.europa.eu/environment/climat/pdf/future_action/part1.pdf

[10] UNFCCC (1992), United Nations Framework Convention on Climate Change, Article 3.1

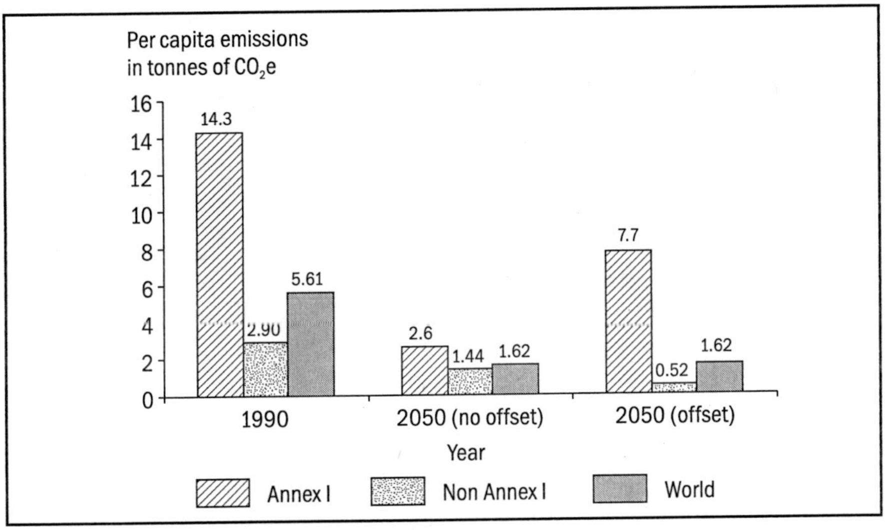

Figure 5 Comparison: per capita emissions
Source WRI (2009); For population: EarthTrends (http://earthtrends.wri.org) Searchable Database Results

measures, the per capita emissions in Annex I would still be above the global average. However, if half of this emission reduction is accomplished through offsets of credits primarily sourced from Non-Annex I, the wide further deepens with Annex I per capita emissions being 15 times that of Non-Annex I. With a six-fold reduction in average per capita emissions of Non-Annex I from 1990, it would be difficult to overcome the developmental imperatives and the poverty alleviation challenges of the developing world. It is for this reason that the Non-Annex I countries led by China opposed a quantified global target combined with a quantified Annex I emission reduction target in the Copenhagen Accord, which only recognizes 2 °C temperature rise limit as a global goal.[11] However, even this was not acceptable to the group of small island states and the least developed countries (LDCs), which suggested a 1.5 °C limit.

Winkler, Vorster, and Marquard (2009) suggest a different approach that would provide a better basis for an equitable burden sharing and also address the distributional issues. It involves a turning around of the formula, wherein the bottom-up approach is taken to quantify the deviation of Non-Annex I countries from the nationally

[11] UNFCCC (2009), Copenhagen Accord document

appropriate mitigation actions reported by them through their national communications or against a registry. This approach would allow for some developmental space to Non-Annex I Parties in undertaking mitigation actions that are most appropriate.

Recent estimates of Climate analytics and Ecofys (Höhne, Schaeffer, and Chen 2009) suggests that the projected global business as usual emissions for 2020 is 57 gigatonnes of CO_2 eq. ($GtCO_2e$) and the emission reduction required to limit temperature rise to 2 °C or 1.5 °C is around 13–17 $GtCO_2e$. Also, Project Catalyst[12] predicts 14 $GtCO_2e$ of emission reductions (for limiting temperature rise to 2 °C) and based on the commitments by developed countries and pledges from developing countries before Copenhagen, estimates that the global emission reduction proposed is 9 $GtCO_2e$ at high end of the proposal range, of which 5 $GtCO_2e$ is from developing countries. It also suggests that 1 $GtCO_2e$ of emissions reduction committed by developed country will be undertaken in developing countries through offset mechanism. Hence, with the current proposals, developing countries are the ones taking the lead. This was also emphasized in Yvo de Boer's final press briefing at the end of Copenhagen, where he stated that the major developing countries with their national action plans attributing 28% emission reduction below a business-as-usual path are more on track in responding to the science than the industrialized world.[13] Considering Winkler's bottom-up approach, the way forward will be to take this 28% emission reduction by Non-Annex I countries as the basis of setting up the target for Annex I countries. This will be a fair approach towards setting up of ambitious Annex I emissions reduction targets and an effective way to fill the mitigation gap.

Nature of compliance mechanism

Any legally binding emissions target is only as good as the compliance mechanism that ensures its implementation. It is for this reason that the issue of an enhanced and stringent compliance mechanism is centrifugal to the present negotiations. Though technical, the

---

[12] Project Catalyst Briefing paper, 2009, Taking stock the emission levels implied by the current proposals for Copenhagen, Project-catalyst accessed at http://www.project-catalyst.info/images/publications/taking_stock.pdf

[13] See <http://unfccc.int/meetings/cop_15/items/5257.php>

issue is politically contentious, as it is linked with other issues in the international climate negotiations like the mitigation debate on emissions reductions target and its compliance and the support in the form of finance and technology.

The recent UNFCCC data on GHG emissions data for Annex I Parties indicates reduced emissions from 1990 to 2004 in 22 Annex I Parties and increased emissions in 19 Parties, with a net emissions reduction of 3.3% during 1990-2004, from 18.6 to 17.9 billion tonnes CO_2e (UNFCCC 2006). For Annex I Parties with economies in transition (EITs), GHG emissions decreased by 36.8% (from 5.6 to 3.5 billion tonnes CO_2 eq.) from 1990 to 2004. This contributed considerably to the overall decrease in GHG emissions from Annex I Parties. However, for the non-EIT Annex I Parties, GHG emissions increased from 1990 to 2004 by 11%, from 11.3 to 14.4 billion tonnes CO_2 eq. From 2000, the emissions increased slightly for both EIT and in non-EIT Parties, as well as for all Annex I Parties taken together. Hence, in relation to the NKP targets for 2008–12 many Parties need to achieve substantial additional emission reductions. Some of the countries that substantially lag behind their KP targets include Japan, Canada, and Norway (Figure 6).

The consequences of Parties' inability to fulfil the commitments on emissions reductions target are not clear. There are, however, certain provisions in the KP and its compliance mechanism, for example, if

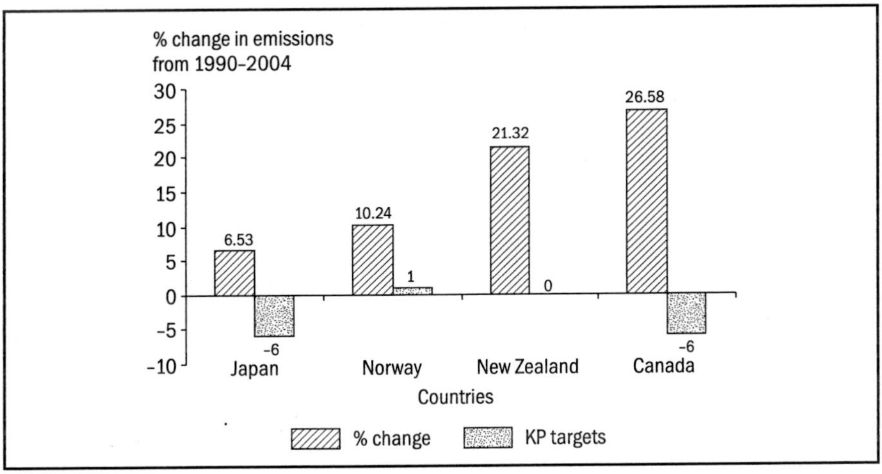

Figure 6 Kyoto progress: Select countries

Data source UNFCCC (2006)

the emissions of a Party have exceeded their assigned amount, the Party is required to make up the difference between its emissions and its assigned amount during the second commitment period, plus an additional deduction of 30%. Hence non-compliance should be restored in the next commitment period at a rate of 1.3. The non-complying country will then have to develop a compliance action plan that provides for action to comply with the quantitative emission commitments of the subsequent commitment period and gives priority to domestic policies and measures. Also, involvement of the country in any form of emissions trading will be suspended.[14] However, non-compliance with emissions targets is not an issue that can come until the end of the commitment period in 2012, as the present system does not allow for a mid- term review of the commitment period. Also, the relevance of this provision (or restored rate) diminishes if a new international framework is eventually agreed upon.

Apart from this, the real issue in climate negotiations revolves round the issue of measurable, reportable, and verifiable (MRV), primarily because of the different reporting requirements of Annex I and non-Annex I countries. Much of the debate focuses on these requirements and guidelines, with different groups suggesting different proposals, including peer review processes to determine progress, a pledge and review mechanism, and so on. Also, the issue of national sovereignty has surfaced, with developing countries reluctant to let their domestic actions reviewed internationally. However, there is lack of attention around the issue of institutionalizing an operational legal process to determine non-compliance and the possible measures to penalize non-compliance, primarily because such a compliance mechanism can also be linked with other trade or border measures outside the UNFCCC alongside internal UNFCCC measures, which can be economically unfavourable.

The Copenhagen Accord, however, did include ambiguous terms such as 'international consultation' and 'analysis' while also providing for defining the guidelines for the same for non-Annex I countries.[15] However, it did not say anything on the compliance mechanism and the associated penalties. Avoiding such an issue, which directly relates to the global emissions reduction goal, and ultimately, to the

---

[14] http://unfccc.int/kyoto_protocol/compliance/items/2875.php

[15] UNFCCC (2009), Copenhagen Accord document

stabilization of GHG concentration, can be detrimental to any progress that leads to combating climate change. Therefore, a way forward will be to construct a robust compliance mechanism that could bring about significant implementation and ensure environmental integrity of pledges and commitments of all Parties.

Economics involved

The bottom line of every debate related to mitigation action is the significant costs required to achieve ambitious abatement levels. The UNFCCC (2007) report on financial flows estimates that global additional investment and financial flows of $200–210 billion will be necessary in 2030 to return to current levels of global emissions.[16] The IPCC indicates that global average macro-economic costs for mitigation are 0.6% of the global gross domestic product (GDP) in the year 2030 and about 1.3% of the GDP in the year 2050.[17] Under the Convention, developed countries shall provide financial resources to meet the agreed full costs incurred by developing countries,[18] as developing countries have developmental priorities to be addressed. However, developed countries see incurring this cost as a disadvantage to their economic competitiveness. Therefore, the key issue under the negotiations includes identifying sources of finance (public/private/market), availability of finance, and the scale at which it is available.

For an effective global response, international climate change negotiations should aim to strike a balance between prescribing emission reduction targets (global or individual) as per its cost effectiveness and also to account for national socio-economic circumstances. The KP offers such cost-effective means of reducing emissions by way of three market-based mechanisms—the European Union Emissions Trading Scheme (EU-ETS), Clean Development mechanism (CDM), and Joint Implementation (JI). The CDM alone, through its more that 2000 registered projects, ensures avoiding more than 1.7 billion tonnes of CO_2 emissions in 2012.[19] However, scaling up of market mechanisms will be required for a robust response.

---

[16] UNFCCC (2007), Background Paper on Investment and Financial Flows to Address Climate Change

[17] IPCC (2007), Climate Change 2007: Synthesis Report

[18] Article 4.3 of the Convention

[19] CDM statistics accessed at http://cdm.unfccc.int/Statistics/index.html

Also, state-of-the-art mitigation technology is a potential solution to climate change. The availability of such environmentally sound technologies (ESTs) will be important for effective global response to climate change. The UNFCCC (2008) report on technology transfer states that only 36% of the CDM projects involve technology transfer[20], which is way below the expected level, given the fact that substantial reductions would take place in developing countries (See above: Binding global emission reduction target: what does it entail?). Therefore, more support in the form of enhanced transfer of ESTs would also be required. Multilateral cooperation on development and transfer of technology would be an important component of any international agreement on climate change.

The current debate, therefore, revolves around the issue of support in the form of finance and technology from developed to developing countries. The Copenhagen Accord provides for resources of about $30 billion for the period 2010–12 with balanced allocation between adaptation and mitigation and also of $100 billion a year by 2020 to address the needs of developing countries[21], but it does not make this operational. Therefore, the way forward to ensure mitigation action in developing countries would be to provide support in terms of technology and finance.

Conclusions

The IPCC *Fourth Assessment Report* states that global GHG emissions have grown since pre-industrial times, with an increase of 70% between 1970 and 2004. With current climate change mitigation policies and related sustainable development practices, global GHG emissions will continue to grow over the next few decades (IPCC 2007). Paradigm shift in mitigation policies and actions is, therefore, required. An important first step would be to steer ahead of the mitigation debate by resolving the issues highlighted above. Some of the key questions that need to be answered in the process through further negotiations and for larger consensus include the following.

- Whether there is a need to revisit the KP or if it is sufficient to have comparable actions from all nations.

---

[20] http://cdm.unfccc.int/Reference/Reports/TTreport/TTrep08.pdf

[21] See UNFCCC (2009), Copenhagen Accord document

- What would be the criteria of graduation or differentiation and when (which commitment period) would it be most appropriate to apply?
- The process to set emissions reduction targets for all nations that will ensure comparable emissions reduction target for all Parties involving the bottom-up approach, wherein the cumulative non-Annex I actions would determine the Annex I targets.
- Institutional arrangements to ensure robust compliance mechanism.
- Support in terms of finance and technology for developing countries.

Resolving these issues would be a prerequisite for a successful and ambitious international framework or agreement on climate change.

References

den Elzen M and Höhne N. 2008
Reductions of greenhouse gas emissions in Annex I and Non-Annex I countries for meeting concentration stabilization targets
Climatic Change **91**:249–274

Gupta S, Tirpak D A, Burger N, Gupta J, Höhne N, Boncheva A I, Kanoan G M, Kolstad C, Kruger J A, Michaelowa A, Murase S, Pershing J, Saijo J, Sari A. 2007
Policies, instruments and co-operative arrangements
In *Climate Change 2007: mitigation*
Cambridge, UK, and New York, USA: Cambridge University Press
[Contribution of Working Group III to the Fourth Assessment Report of the Intergovernmental Panel on Climate Change [B Metz, O R Davidson, P R Bosch, R Dave, L A Meyer (eds)]

Höhne N, Schaeffer M, and Chen C. 2009
Deal – how to close the gap [Briefing paper]
Climate Analytics and Ecofys
Details available at <http://www.climateactiontracker.org/briefing_paper.pdf>

IPCC (Intergovernmental Panel on Climate Change. 2007
Summary for policy-makers
In *Climate Change 2007: mitigation*
Cambridge, UK, and New York, USA: Cambridge University Press
[Contribution of Working Group III to the Fourth Assessment Report of the Intergovernmental Panel on Climate Change [B. Metz, O.R. Davidson, P.R. Bosch, R. Dave, L.A. Meyer (eds)]

UNFCCC (United Nations Framework Convention on Climate Change). 2006
GHG Data 2006: highlights from greenhouse gas emissions data for 1990-2004 for Annex I Parties
Details available at <http://unfccc.int/files/essential_background/background_publications_htmlpdf/application/pdf/ghg_table_06.pdf>, last accesssed on 27 January 2010

UNFCCC (United Nations Framework Convention on Climate Change). 2007
Background Paper on Investment and Financial Flows to Address Climate Change, Bonn
Details available at http://unfccc.int/files/cooperation_and_support/financial_mechanism/application/pdf/background_paper.pdf > last accessed on 21 January 2010

UNFCCC (United Nations Framework Convention on Climate Change). 2008
Analysis of Technology Transfer in CDM Projects, Bonn
Details available at <http://cdm.unfccc.int/Reference/Reports/TTreport/TTrep08.pdf> last accessed on 21 January 2010

Winkler H, Vorster S, and Marquard A. 2009
Who picks up the remainder?
Mitigation in Developed and Developing Countries 9 (6): 634-651 (18)

WRI (World Resources Institute). 2009
Climate Analysis Indicators Tool (CAIT) Version 6.0
Washington, DC: WRI

WRI (World Resources Institute). 2010
Climate Analysis Indicators Tool (CAIT) Version 7.0
Washington, DC: WRI

Bionotes

Björn Stigson, President, World Business Council for Sustainable Development (WBCSD) has had extensive experience in international business. He began his career as financial analyst with the Swedish Kockums Group. From 1971 to 1982, he worked for ESAB, the international supplier of equipment for welding, in different positions responsible for Finance, Operations and Marketing. In 1983, he became President and CEO of the Fläkt Group, a company listed on the Stockholm Stock Exchange and the world leader in environmental control technology. In 1991, he became Executive Vice-President and a member of ABB Asea Brown Boveri's Executive Management Group. In 1995, he was appointed President of the WBCSD in Geneva.

The WBCSD is a unique, CEO-led, global association of some 200 companies dealing exclusively with business and sustainable development. The Council provides a platform for companies to explore sustainable development, share knowledge, experiences and best practices, and to advocate business positions on these issues in a variety of forums, working with governments and non-governmental and intergovernmental organizations.

Farhana Yamin is an international lawyer specializing in climate change law and policy and now works with the Special Climate Change Initiative at the Children's Investment Fund Foundation. She was Director of the Climate Change and Energy Programme at the Foundation for International Environmental Law and Development and provided legal advice and assistance to small island states on the climate negotiations from 1992–2002, including negotiations relating to the Kyoto Protocol and Marrakech Accords. She was the Director of the consortia that advised the European Commission on the legal and policy framework of the European emissions trading scheme. As a Research Fellow at the Institute of Development Studies, University of Sussex, during 2003–09, she directed a wide range of policy relevant research and capacity development relating to carbon markets and linkages between development and climate policy with particular reference to Brazil, China, India and South Africa (the BASIC Project). She has been a lead author for the Working Group I and II of the Intergovernmental Panel on Climate Change since 1994 and contributed to the Second, Third and Fourth IPCC Assessment Reports.

Jessica Ayers is a PhD candidate at the Development Studies Institute, London School of Economics (LSE), and a research consultant for the International Institute for Environment and Development (IIED). Her work focuses on the governance of adaptation to climate change, specifically the relationship between community-based adaptation, and wider climate change and development governance structures at the national and international level. She also acts as an independent consultant for a variety of development and environment organizations working to improve policy and action on climate change adaptation, and has published numerous articles and chapters in peer reviewed journals, books, and popular publications on these issues. She is currently working for the UNDP in Nepal on designing the National Adaptation Programme of Action.

Jiahua Pan is currently Director, Institute for Urban and Environmental Studies, Chinese Academy of Social Sciences (CASS), and Professor of Economics at CASS Graduate School. Received his PhD at Cambridge University in 1992. Worked for the UNDP Beijing Office as an advisor on environment and development; lead author of the IPCC Working Group Third and Fourth Assessment Reports on Mitigation; Member of China National Expert Panel on Climate Change; Advisor to the Ministry of Environment Protection. Vice-president of the Chinese Society of Ecological Economists. Co-editor of *Climate Change 2007: mitigation*, published by Cambridge University Press, and author of over 200 papers, articles and books in both English and Chinese.

Joanna Depledge is an Affiliated Lecturer at the Department of Politics and International Studies, Cambridge University (UK). She has published several papers on climate change politics and the international climate change negotiations, and is a regular contributor to the journal *Environmental Policy and Law* on a range of environmental issues. Joanna Depledge is the co-author of *The International Climate Change Regime: a guide to rules, institutions and procedures* (2004) and author of *The Organization of global negotiations: constructing the climate change regime* (2005). She has worked for the UN Climate Change Secretariat, and also as writer/ editor for the *Earth Negotiations* Bulletin.

Leena Srivastava is currently the Executive Director, TERI, New Delhi. In her 26+ years of experience at TERI, she has worked on a range of issues covering energy and environment policy/planning, energy economics and climate change. Dr Srivastava held additional charge as Dean, Faculty of Policy and Planning, TERI University, from June 2000 to June 2008. She has a PhD in Energy Economics from the Indian Institute of Science, Bangalore, India, and has a number of publications to her credit. She is on the editorial boards of various international journals dealing with energy and environment issues.

Dr Srivastava is a member of the Advisory Group on Energy and Climate of the UN Secretary General; Member, International Advisory Panel, Global Carbon Capture and Storage (CCS) Institute, Member, ESMAP Expert Panel on 'Sustainable Energy Supply, Poverty Reduction and Climate Change', an Independent Director on Reliance Infrastructure Ltd; Member of the International Public Policy Advisory Board (IPPAB), the Coca Cola Company and Member of the Foresight Advisory Council of Suez Environment. She was a member of the Expert Committee to formulate India's Energy Policy, Planning Commission, Government of India, and Member, National Security Advisory Board, Government of India. She serves on the research advisory councils of various academic institutions of international repute. She was a Co-ordinating Lead Author for Working Group III of the *Third Assessment Report* of the Intergovernmental Panel on Climate Change (IPCC) and cross-cutting theme anchor on 'Sustainable Development' for the *Fourth Assessment Report* of the IPCC.

Neha Pahuja is a Research Associate at the Centre for Global Environment Research, TERI. She is a management graduate in environment and development studies. Her work focuses on the international climate policy and negotiations, wherein she has worked on quantifying the principle of historic responsibility and bringing developing country perspectives in range of global governance issues such as international financial institutions, MRV, undertaking nationally appropriate mitigation action, global registry, and so on. She is also working on the implication of the new international regime on India and the domestic policies. Prior to this, she worked with a leading carbon advisory firm in India on business responses to climate change and global carbon markets.

Prodipto Ghosh, Distinguished Fellow, TERI; Member, Prime Minister's Council on Climate Change, and former Secretary, Environment & Forest in the Union Government from September 2003 to May 2007. He specializes in the interface of science, economics, and public policy. He has a PhD in Economics and Policy Analysis from the Carnegie-Mellon University, a BTech in Chemical Engineering from IIT Delhi, and was a member of the Indian Administrative Service (IAS) from July 1969 to May 2007. He works on climate change and sustainable development policy.

Rajendra K Pachauri is a renowned economist and leading environmentalist, who heads The Energy and Resources Institute (TERI) as Director-General. He has served as the Chairman of the Intergovernmental Panel on Climate Change (IPCC) since 2002. He has been active in several international forums dealing with the subject of climate change and its policy dimensions. Dr Pachauri completed his MS degree in Industrial Engineering from North Carolina State University, Raleigh, in 1972, as well as a joint PhD in Industrial Engineering and Economics in 1974. He served as Assistant Professor (1974/75) and Visiting Faculty Member (Summer 1976 and 1977) in the Department of Economics and Business at NC State. On his return to India, he joined the Administrative Staff College of India, Hyderabad, as Member Senior Faculty (1975–79) and went on to become Director, Consulting and Applied Research Division (July 1979–March 1981). He joined TERI as Director in April 1981.

To acknowledge his immense contribution to the field of environment, Dr Pachauri was awarded the Padma Vibhushan, one of India's highest civilian awards by the Government of India, in 2007. He is currently a member of 16 international committees and boards. Dr Pachauri provides expert advice to both the central and individual state governments of India. He directly advises the Prime Minister of India as a leading member of the Prime Minister's Advisory Council on Climate Change and as a contributing member of government bodies in the areas of power and energy, petroleum and natural gas, industry and public enterprises and environment.

Dr Pachauri is the author of 23 books and over a hundred papers and articles. He contributes regularly to international media outlets and has given presentations in over 50 countries worldwide on the subject of climate change.

Richard Sandor is Chairman of the Chicago Climate Exchange, the world's first and North America's only voluntary, legally binding integrated greenhouse gas emissions reduction, registry and trading system and Chicago Climate Futures Exchange. Dr Sandor is also Executive Chairman of Climate Exchange plc. (CLE.L), an AIM-listed company which owns the CCX and CCFE and the European Climate Exchange, Europe's leading exchange operating in the European Union Emissions Trading Scheme. Dr Sandor is a research professor at the Kellogg Graduate School of Management at Northwestern University and a Member of the International Advisory Council of Guanghua School of Management at Peking University. While on sabbatical from the University of California, Berkeley in the early 1970s he served as vice-president and chief economist of the Chicago Board of Trade. It was at that time that he earned the reputation as the principal architect of the interest-rate futures market. Richard Sandor was honored by the City of Chicago and the Chicago Board of Trade for his contribution to the creation of financial futures and his universal recognition as the 'father of financial futures'. In October 2007, Dr Sandor was honoured as one of TIME Magazine's 'Heroes of the Environment' for his work as the 'Father of Carbon of Trading'. Dr Sandor received his Bachelor of Arts degree from the City University of New York, Brooklyn College, and earned his PhD in Economics from the University of Minnesota in 1967.

Saleemul Huq is a recognized international expert on adaptation funding and community-based adaptation. He has broad experience in providing climate change and development advice to political leaders, in establishing and developing research centres in climate change and adaptation that inform the policy debate, and in working in the most vulnerable countries (in particular Bangladesh). He is familiar with the UNFCCC negotiating process and provides technical advice to a number of national negotiating teams. As a Member of the Intergovernmental Panel on Climate Change, Dr Huq received the Nobel Peace Prize in 2007. His academic qualifications are as below.

- PhD Botany, Imperial College, London University 1979
- DIC Botany Imperial College, University of London 1978
- BSc (Hons) Botany Imperial College, University of London 1975
- 2001 to present: Climate Change Group, IIED, London, UK
- 2000 to 2001: Imperial College, London, UK
- 1985 to 2000: Bangladesh Centre for Advanced Studies, Dhaka, Bangladesh
- 1979 to 1985: Dhaka University, Dhaka, Bangladesh

Thomas J Wilbanks is a Corporate Research Fellow at the Oak Ridge National Laboratory. He conducts research on such issues as sustainable development, responses to environmental hazards and changes, and the role of geographical scale in these regards. Co-edited recent books include *Global Change and Local Places* (2003), *Geographical Dimensions of Terrorism* (2003), and *Bridging Scales and Knowledge Systems: linking global science and local knowledge* (2006). He played roles in the first US National Assessment of Possible Consequences of Climate Variability and Change (1997–2000); the Intergovernmental Panel on Climate Change (IPCC) Working Group II (Impacts, Adaptation, and Vulnerability) *Third Assessment Report*; and aspects of the UNEP. Millennium Ecosystem Assessment related to issues of geographic scale and regional and local assessments. He served as Coordinating lead author for the IPCC's *Fourth Assessment Report*, Working Group II, Chapter 7: 'Industry, Settlement, and Society' and as a lead author of the Working Group II Technical Summary and Summary for Policy-makers

He has led several 'synthesis and assessment' reports for the US Climate Change Science Program (CCSP), including summaries of current knowledge about impact and resilience issues for human settlements in the US and for energy production and use in the US Wilbanks is Chair of NRC's standing Committee on Human Dimensions of Global Change, and he serves on a number of other NRC committees and panels, including the NAS/NRC Committee on America's Climate Choices and serving as vice-chair of the adaptation panel for that study.

Ying Chen, Senior Research Fellow, Deputy Director and Secretary-General of Research Center for Sustainable Development (RCSD), Chinese Academy of Social Sciences (CASS) and Associate Professor at CASS Graduate School. Her research interests include international environmental governance, climate policy and sustainable development. She has published two books (co-author), over 50 papers in academic books and journals, and a number of reports, as well as analyses and articles in newspapers.

Index